Building a High-Tech Alarm System with Raspberry Pi
(Second Edition)

William Pretty

Copyright © 2024 by William Pretty

All rights reserved. This book or any of its portion may not be reproduced or transmitted in any means, electronic or mechanical, including recording, photocopying, or by any information storage and retrieval system, without the prior written permission of the copyright holder except in the case of brief quotations embodied in critical reviews and other noncommercial uses permitted by copyright law.

Printed in the United States of America
Library of Congress Control Number: 2024923320
ISBN: Softcover   979-8-89518-441-7
           e-Book   979-8-89518-442-4
Published by: WP Lighthouse
Publication Date: 10/25/2024

To buy a copy of this book, please contact:
WP Lighthouse
Phone: +1-888-668-2459
support@wplighthouse.com
wplighthouse.com

Contents

Chapter 1: Introduction to Alarm Systems 1
    Motion Detectors: 3
    Glass Break Sensor: 4
    Fire Alarm Sensors: 5
    Access Control: 11

Chapter 2: Hardware 14
    Voice Output: 29

Chapter 4: Software 35

Chapter 5: Printed Circuit Board 57
    Printed Circuit Board 62

Chapter 6: Alarm System Wiring 72
    Testing the Harnesses 75
    Wiring the Sensors 78

Chapter 7: Planning your Alarm System 87
    Step 1 – The Walk-about 88
    Typical Four Bedroom House 89
    Commercial Office Space 96
    Laboratory 98

Chapter 8: Future Enhancements 101
    CAD Software: 110
    Smoke Alarms, Solenoid Locks, Suppliers: 111

Chapter 9: Adding more Inputs and Outputs 112
    Trouble Shooting the System: 115

Chapter 1:

Introduction to Alarm Systems:

In this chapter we will discuss the basic components of any alarm system. All alarm systems have two basic functions. First, they monitor their environment looking for a change such as a door or window opening or someone moving about in the room. The second function of the system is to alert the human to this change. Our alarm system uses a scanning type software to detect intruders. We will use the 'standard' guard dog as an analogy. In a scanned type of system, the guard dog paces back and forth at the fence looking out for either an intruder or someone that it recognizes. In our design, if you have an alarm key, you can disarm the system and enter. In an interrupt driven system, the dog is asleep until it hears an intruder (or you). It then wakes up and deals with the situation. I have chosen the scanning method because in my opinion the software is easier to write and explain. It can scan all eight zones in about one second.

You don't have to be an electrical engineer to install an alarm system, just a good carpenter, painter, and plasterer! I'm not by the way so I'll leave hiding wires up to you.

Also, because our alarm system runs on 12 volts, you don't have to be a licensed electrician to install it. If you can plug in a wall adapter, you can build and test this alarm system.

Alarm System Sensors:

Door / Window Contact:

The simplest and one of the most common sensors is the door / window contact. This sensor consists of a magnet which is installed on the moving part of the door or window. This magnet holds a switch closed. The switch portion of the sensor is attached to the door or window frame.

Figure 2 shows what is inside a typical sensor of this type.

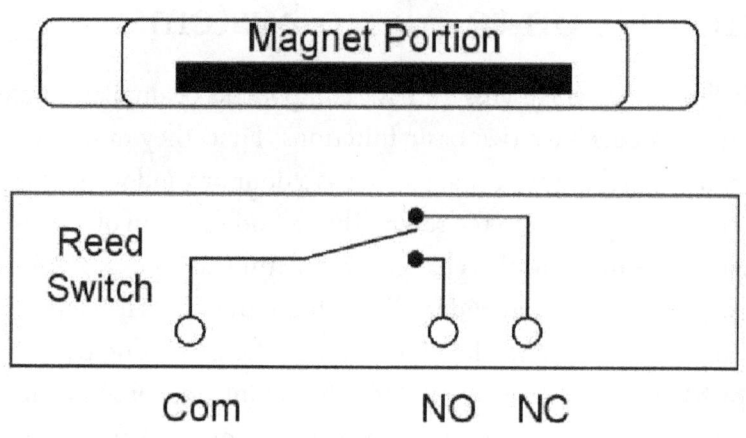

Figure 1-1. Door / Window Contact Schematic

Figure 1-2. Door / Window Contact

Motion Detectors:

The next most common sensor is the PIR or Passive Infra-Red detector. This sensor measures the ambient temperature of the room and waits for a change in the ambient. Often called a 'blip'.

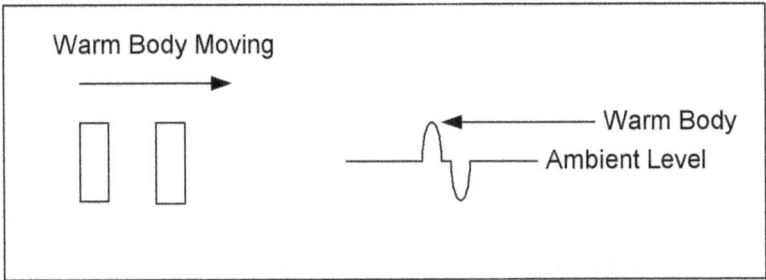

Figure 1-3. How PIR Sensor works

Simple PIR sensors tend to be fooled by large pets (like your guard dog). For that reason, they have a "pet" setting which ignores any object less than 30 pounds, which is moving close to the floor. More modern (expensive) sensors also have a mmWave sensor built in which tends to reduce false alarms and makes the sensor harder to fool. The dual sensor is about three times the price of a simple PIR detector and communicates with the panel with the same contact switch arrangement.

Figure 1-4. PIR Motion Sensor

Glass Break Sensor:

Another type of sensor is the glass break sensor. This type of sensor is commonly used by shop owners to help detect vandalism. This sensor uses a microphone to 'listen' for the sound of breaking glass.

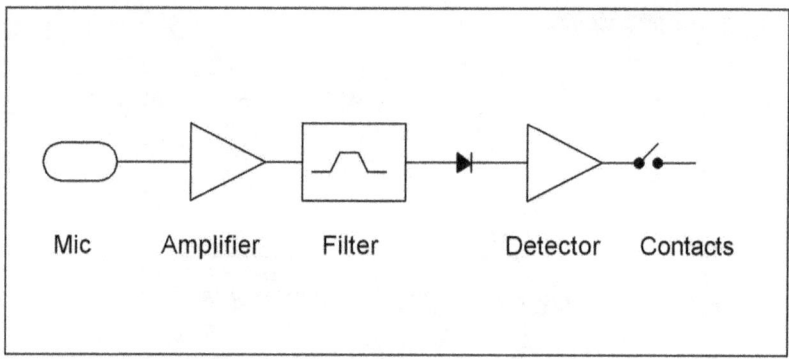

Figure 1-5. How Glass Break Sensor works

The system consists of a sensitive microphone, an amplifier, and a filter (usually digital signal processing). The output of this filter is connected to a detector circuit which activates the alarm system contacts when the sound of breaking glass is 'heard' by the microphone.

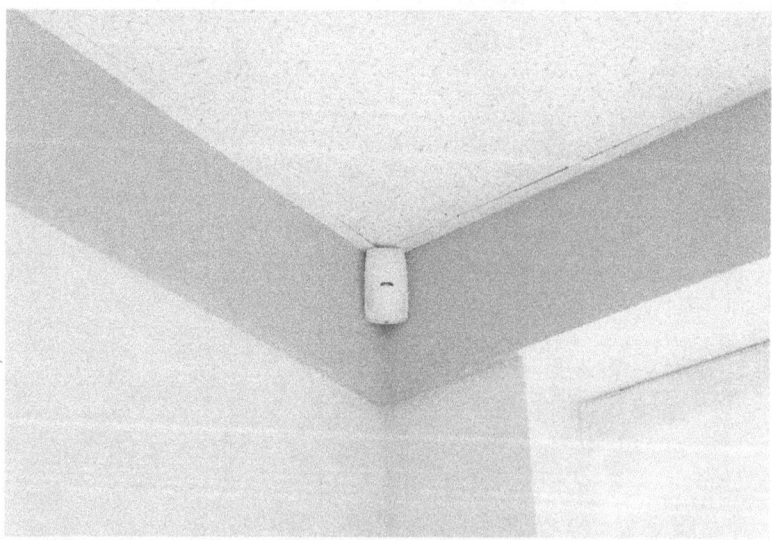

Figure 1-6. Typical Glass Break Sensor

Fire Alarm Sensors:

Heat Detectors:

There are two basic types of heat detectors, mechanical and electronic. The electronic type of detector uses a thermistor as the sense element. A thermistor is type of resistor that changes its value based on the ambient temperature. In practice two thermistors are used. One resistor is exposed to the ambient air and the other is partially sealed from the surrounding air. A fast rise in the air temperature, for example from an open flame, is sensed by the resistor which is exposed to the surrounding air. The resistor changes its value, and an alarm is triggered. A slow rise in temperature, from a smoldering fire, is sensed by both resistors and again an alarm is triggered.

There are two types of mechanical heat detector, bimetallic and pneumatic.

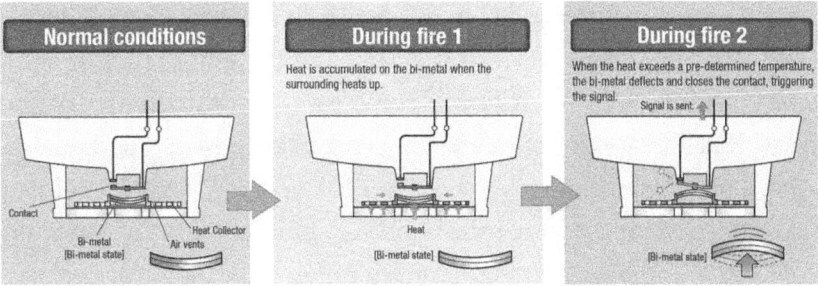

Figure 1-7. Bimetallic Heat detector © Hochiki

A bimetallic heat detector contains a metallic strip consisting of two different metals bonded together. When heat reaches the strip, it bends until the contacts are closed and an alarm condition is signaled.

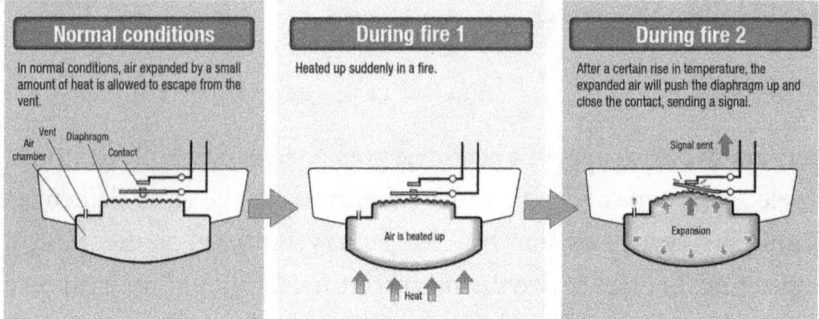

Figure 1-8. Pneumatic Heat sensor © Hochiki

Another type of heat detector is the fixed temperature heat detector. As the name suggests, the sensor is set to trigger at a specific temperature. The trigger temperature is usually marked on the outside of the device. The device shown below is a combination of ROR and fixed temperature device.

ROR heat detectors may not respond to smoldering fires. For that reason, a fixed element is also included in the device, in order to detect this type of fire. The small metal disk in the center of the unit is part of fixed element. In the bottom view, you can see a tube in the center of the device. This tube is filled with a wax like substance which is designed to melt at a preset temperature. In this case 136F or 67C. The small pointed tube to the right of the large tube is the vent for the rate of rise detector.

Figure 1-9. Fixed Temperature / ROR Alarm

Figure 1-10. Bottom View

Heat detectors are useful in areas such a garages, kitchens, or workshops. Where a certain amount of smoke or fumes is 'normal' but would trigger a smoke alarm. This is due to dust or smoke particles in the air.

Fire Alarm Sensors:

Smoke Alarms:

There are three types of smoke alarms. These are photoelectric, ionization and a combination of both. The ionization type smoke detector uses a small radioactive source, usually americium-241. The detector consists of two positive and negative charged electrodes inside an ionization chamber.

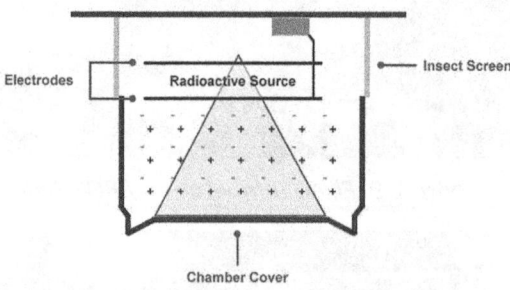

Figure 1-11. Ionization Smoke Detector © Safelincs

The radioactive alpha particles cause a small current to flow between the two electrodes. This current is sensed by the electronics of the detector. If smoke particles enter the chamber thru the bug screen, then the current flow is interrupted.

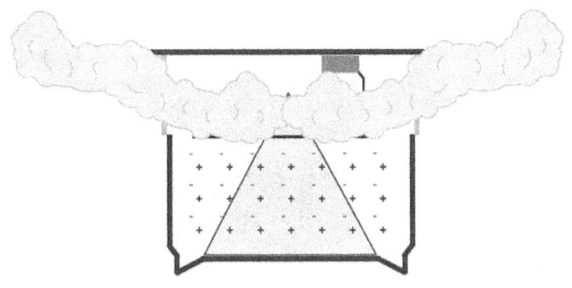

Figure 1-12. Smoke enters the detector © Safelincs

Once enough smoke has entered the chamber, the current change is detected, and an alarm condition is sent to the alarm panel. This type of detector is called a four-wire detector, because two wires are used to power the electronics and two are used for the alarm contacts.

The other type of smoke detector is the photoelectric based detector. There are two types of photoelectric smoke detector. The first type we will be discussing is the scattering type detector.

There is no direct path between the light source and the light receiver. This is due to a series of baffles inside the detection chamber.

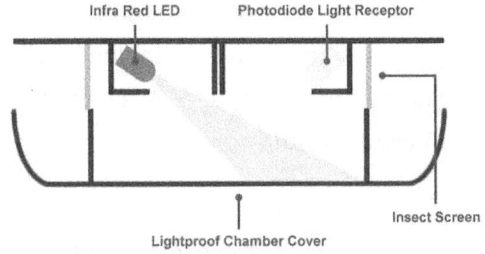

Figure 1-13. Photoelectric Detector © Safelincs

These baffles are arranged is such a way as to allow smoke to enter the chamber, while keeping light out.

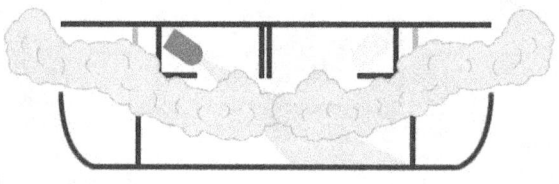

Figure 1-14. Smoke Enters Chamber © Safelincs

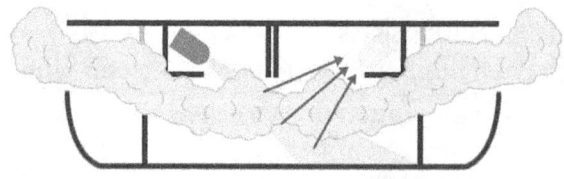

Figure 1-15. Light Scattered by Smoke © Safelincs

Smoke enters the chamber and 'scatters' the light from the transmitter. This light is then detected by the photo receiver. Once a sufficient mount light is detected, an alarm condition is sensed.

The second type of photoelectric detector allows smoke to obscure the infra-red beam. When this happens an alarm sounds.

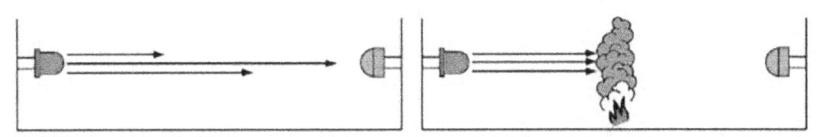

Figure 1-16. Light Obscured by Smoke

It is worth mentioning that these types of optical detector use infra-red LEDs, so as to reduce false alarms due to ambient light entering the chamber.

Access Control:

The simplest type of access control is the key switch, and it is the type I am using in the design of this alarm system. The switch that I am using takes a round key and is an on-off-on switch. When the switch is rotated left or right, the common contact is connected to one of the two other terminals. The key is only removable in the center or off position.

Figure 1-17. Key Switch©Digikey

A keypad can also be connected to the alarm panel. The pad is 4 x 3 matrix of keys. This matrix is scanned by a microprocessor and an ASCII character is sent to the panel when a button is pushed. In some cases, the microprocessor is part of the keypad, in others it is on the main board of the alarm system. For our system, we would have to use an Arduino to scan the keys. I decided to keep the design simple and use a key switch as the standard access control method.

Figure 1-18. Alarm panel Keypad

A good alternative to the keypad is an RFID card reader. This reader is easily connected to the Raspberry Pi via one of its USB ports. The card reader looks like a keyboard to the Pi and sends an ASCII string to the system. The string consists of the serial number programed onto the card, followed by a carriage return. In this way it mimics the user typing on a keyboard. The system then checks a database (text file) of valid serial numbers. If the number is valid it takes the appropriate action.

Figure 1-19. RFID Card Reader

References:

https://www.safelincs.co.uk/smoke-alarm-types-ionisation-alarms-overview/

https://www.electronicsforu.com/technology-trends/smoke-detectors-fire-alarms-guide

https://medium.com/@chuanjerlim/confession-of-a-photoelectric-smoke-alarm-3be8bbd65af9

https://www.hochikieurope.com/products

https://www.hochikiamerica.com/

Chapter 2:

Hardware

The main component of the alarm system is the comparator. The comparator we are using is an LM2903P. The LM2903P is a dual comparator in an 8-pin dip package with open drain outputs.

A comparator is a special purpose amplifier which has two analog inputs, called V+ and V-. The comparator outputs a digital signal Vo, which indicates which input is larger than the other.

The output is the following: Vo = 1 if V+ > V-

Vo = 0 if V- > V+

The following figure shows the zone circuit in 'normal' operation.

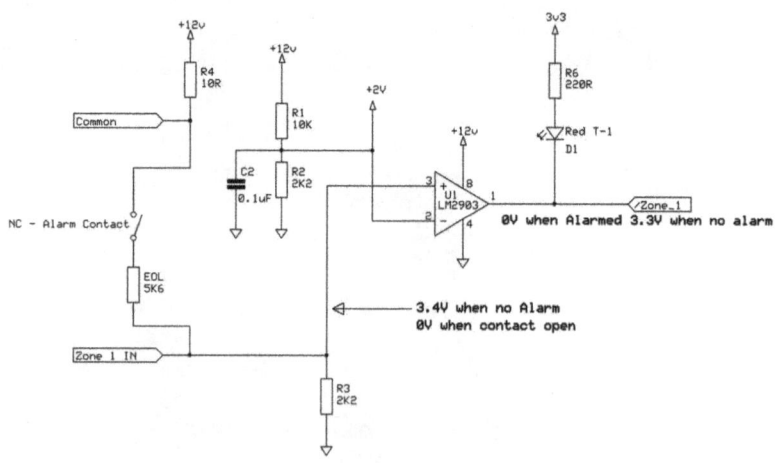

Figure 2-1. Open Detection

Resistors R1 and R2 form a voltage divider which applies approximately two volts to the inverting input of the comparator. R3, R4 and the end of line resistor (EOL) form another voltage divider. This divider is connected to the noninverting input of the comparator. When the alarm contact is closed (no alarm), approximately three point four volts will be measured at the noninverting input to the comparator. This means that the noninverting input is higher than the inverting one. This causes the output to remain high. (The output of the LM2903P has an open drain output, so the internal GPIO pull up resistor can cause the input to be pulled high.)

When an alarm is triggered, or the wire is cut, R3 pulls the input low. Now the noninverting input is lower than the inverting input. The output transistor is turned on and the output is pulled low.

The following figure shows what happens if a burglar attempts to short the alarm contacts. The LM2903P is a dual comparator so, we will use the second half to sense this condition. The following figure shows the result of this condition.

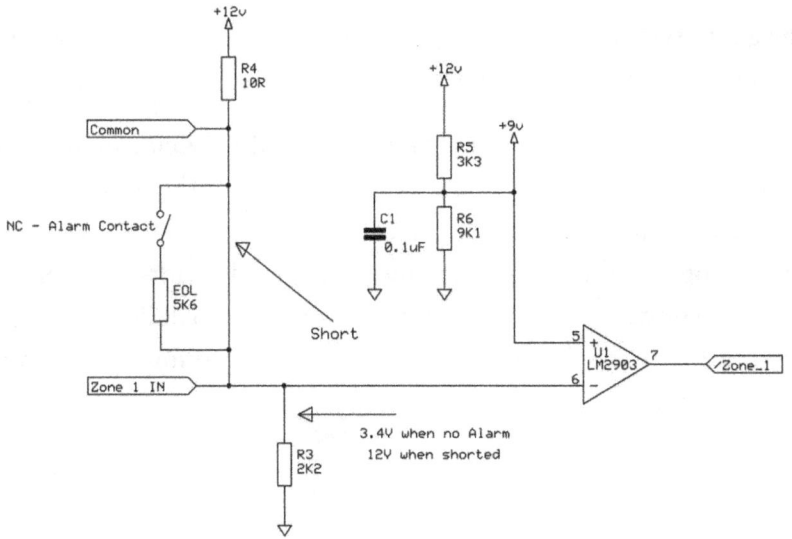

Figure 2-2. Short Detection

In this case resistors R5 and R6 form a voltage divider which is connected to the noninverting input of the comparator. This results in approximately nine volts being applied to the input. If there is no short, then there is three point four volts applied to the inverting input. This means that the inverting input is lower than the noninverting input. So, the output is turned off and the internal pullup resistor pulls the input high. If a burglar attempts to short out the switch, they will also short out the end of line resistor. This results in twelve volts being applied to the inverting input. Now the inverting input is higher than the noninverting input. This causes the output of the comparator to go low and an alarm condition occurs.

The LM2903P is an open drain output device. This allows us to connect the two outputs together in a wired 'OR' configuration. This means that if either a short circuit OR an open one occurs, the input to the Raspberry Pi will be pulled low. The cathode of a red LED is also connected to the outputs. The anode is connected to three point three volts thru a series resistor. When either output goes low, the LED turns on to indicate that there is an alarm condition on that zone.

The alarm system consists of eight zones, so the circuitry we have just discussed is repeated eight times.

There are two special zones called 'fire zones' these zones are intended to be connected to a two-wire heat rise detector or the normally open contacts of a four-wire smoke detector. The contacts on this type of detector operate the opposite to other four wire detectors. The contacts of a two-wire detector are normally open and close to indicate an alarm. The LM2903P is a dual comparator, so the alarm system can monitor two fire zones.

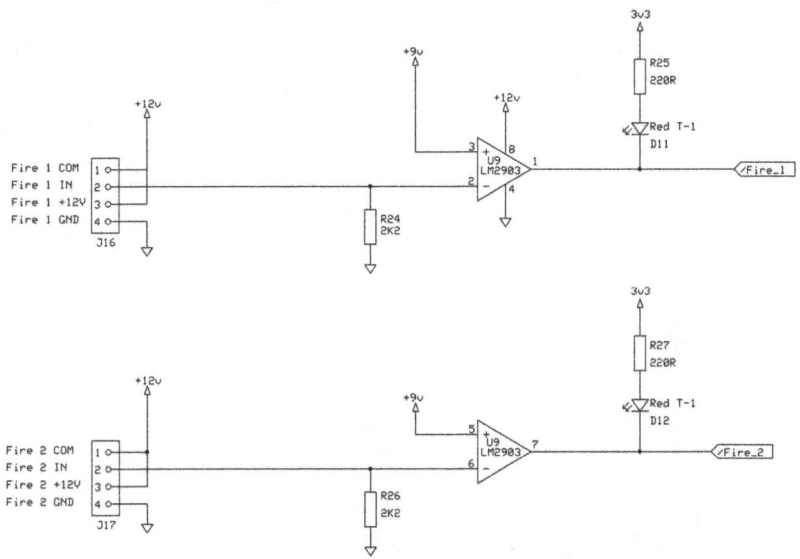

Figure 2-3. Fire Zones

One terminal of the fire zone 1 detector is connected to pin 1 of J16. The other terminal is connected to pin 2.

One terminal of the fire zone 2 detector is connected to pin 1 of J17. The other terminal is connected to pin 2 of J17.

The contacts are normally open, so resistors R24 and R26 pull the inverting input to ground. This makes the inverting input lower than the noninverting input which is connected to the 9V reference. The output transistor is turned off and the internal pullup resistor pulls the output high. (Vo = 1 if V+ > V-)

When excessive heat or smoke is detected, the contacts close, and the inverting input is pulled up to 12V. This makes the inverting input higher than the noninverting input. The output transistor is turned on and the output is pulled low. (Vo = 0 if V- > V+)

The next zone is a special zone called "Tamper". This zone input is connected to a normally open limit switch. The switch is connected to ground and is held closed by the lid of the alarm system. When the system is armed and the cover is removed, the internal pullup resistor pulls the GPIO pin high, and an alarm is triggered.

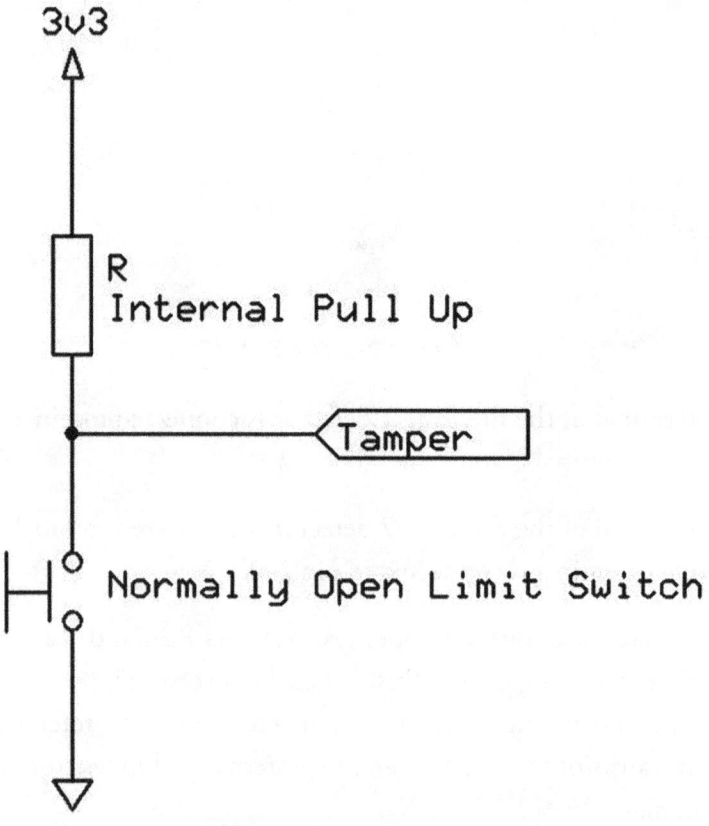

Figure 2-4. Tamper Switch Circuit

External power is supplied to the board via connectors J5 and J6. A twelve-volt four-amp power cube is connected to J6, and a 12-volt SLA (Sealed Lead Acid) battery is connected to J5. These two sources are connected together by power diodes D1 and D2. The board derives its twelve-volt power from F2, which is a three-amp PTC fuse. A PTC fuse is like a solid-state circuit breaker. The letters PTC stand for Positive Temperature Coefficient. This means that as more current flows thru the device it heats up and the resistance goes up, until finally the resistance goes very high. This is called the avalanche point. In this case the point is three amperes.

You may be wondering why I specified a four-amp power supply. This is because I always try to over specify the capacity of the power supply so as not to over tax the supply by running it at its maximum capacity. As for the battery. If you want the system to be able to run for four hours on battery power, then you will need a twelve-volt, twelve-amp hour battery as a minimum. SLA batteries are usually discharged at a one third-C rate. That is, one third of their capacity. Unlike some commercial alarm panels, our design does not include a battery charging circuit. Therefore, the battery should be checked from time to time and recharged if necessary, using a constant voltage charger. Such as an automotive battery charger.

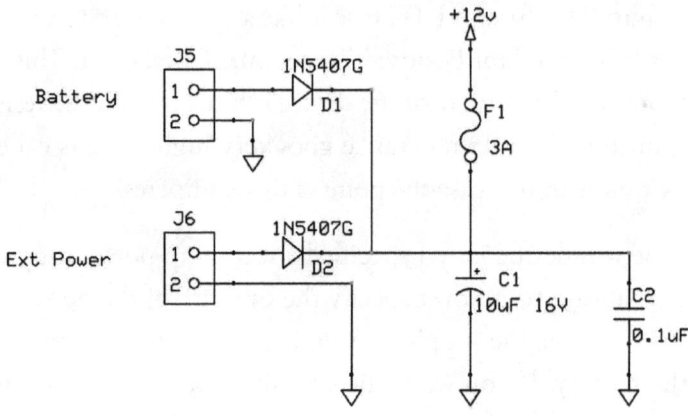

Figure 2-5. System Power

A siren which draws up to two amps DC can be connected to J4. An opto isolator which contains a power MOSFET as its output is used to switch the siren. Note that the MOSFET can only switch DC loads. That is why J4 is connected to the onboard twelve-volt supply. The MOSFET is only protected by the three-amp fuse, so care should be taken when selecting the siren or other load. If more than two amps DC or an AC load such as lighting is connected to the output, then a twelve-volt mechanical relay should be connected to the output.

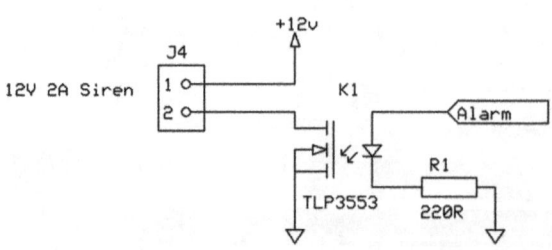

Figure 2-6. Alarm Output

Chapter 3: Human Interface

LCD Display:

In this chapter we will discussing how the alarm panel communicates with the natural world. First, we will discuss the 2x16 LCD display. This display can display 2 rows of 16 characters. There are two methods of sending data to the display. We will be using the type with a serial interface because it is easy to connect to the Raspberry Pi and easy to program.

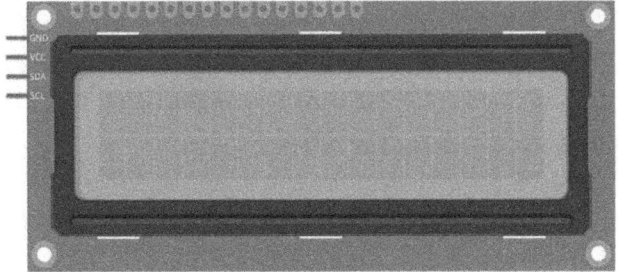

Figure 3-1. 2x16 LCD Front View

Figure 3-2. Rear view

The pinout of J3 on the mother board is identical to the one on the adapter board soldered to the back of the display. If we consider the pin of the adapter closest top edge to be pin 1, then pin 1 is connected to pin 1 (Gnd) of J3 on the alarm panel motherboard. Pin 2 is connected to pin 2 (+5V) of J3. Pin 3 is connected to pin 3 (SDA) of J3 and Pin 4 is connected to pin 4 (SCL) of J3.

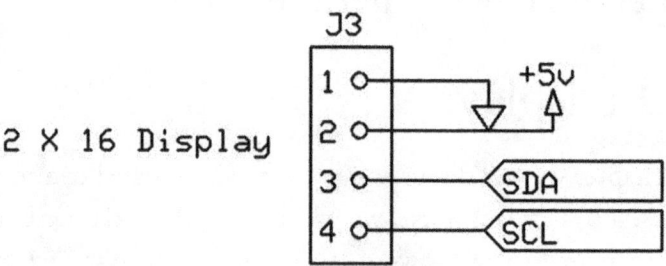

Figure 3-3. Motherboard J3

The first thing we must do is to enable the I2C interface on the Raspberry Pi. The easiest way to do this is to configure a development system with an HDMI display, keyboard, and mouse. My system is running the latest version of the OS (Buster). Click on the raspberry icon on the top left of the screen and you will see the following menu.

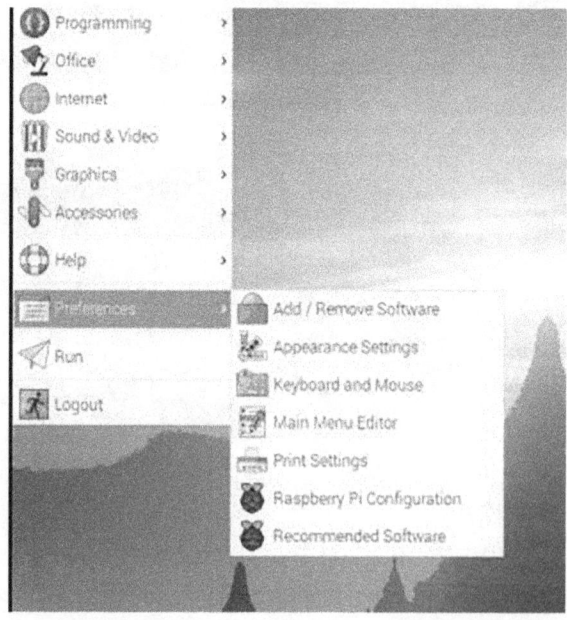

Figure 3-4. Raspberry Pi Configuration

Next, select "Raspberry Pi Configuration" and you will see the following pop-up window.

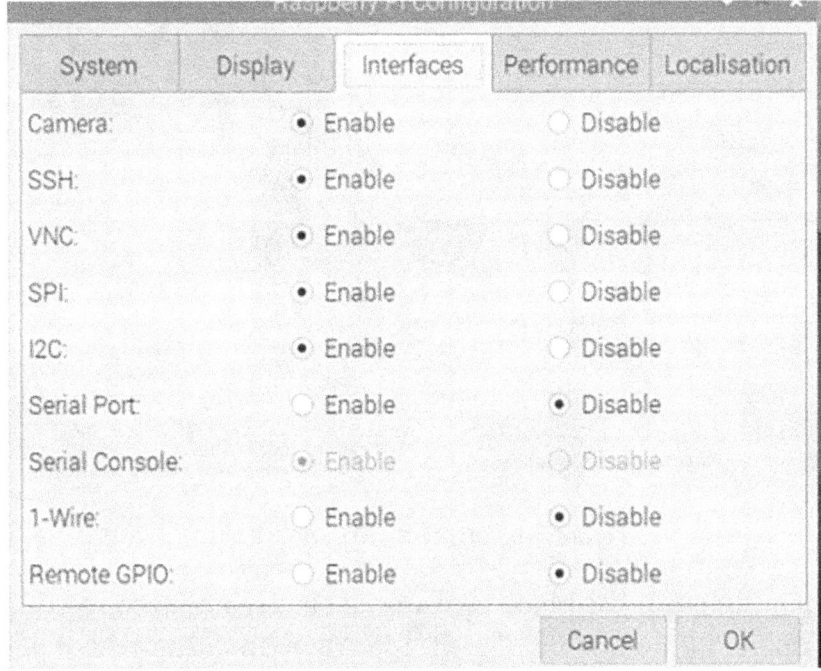

Figure 3-5. Interfaces

Select the "Interfaces" tab and use the radio button to make sure that the I2C interface is enabled. We will have to reboot the system for any changes to take effect. So now is a good time to power down the Raspberry Pi and connect the display to the Raspberry Pi.

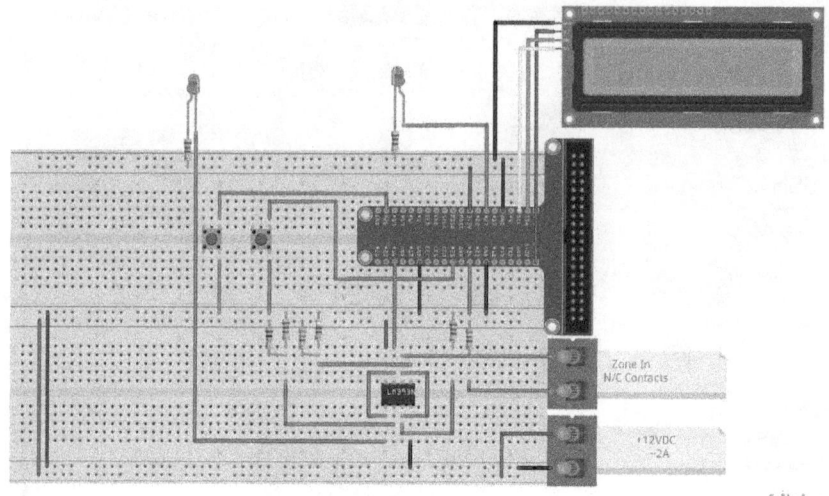

Figure 3-6. Raspberry Pi Connections

You will notice that I have used a breakout adapter to connect the Raspberry Pi to my breadboard. I highly recommend this. It will make your building and test a lot easier. As a bonus, it comes with a 40-conductor ribbon cable which we can eventually use to connect the Raspberry Pi to the motherboard of the alarm system.

It is now time to power up the system and start writing our first python script. Before we can do that, we must install a library and some tools. The first thing we must do is make sure that PIP is installed. We do this with the following command:

sudo apt-get **install** python3-pip

This will install PIP3, if it was not already included with Buster. Once we have PIP3 installed, we can install the serial LCD library we will be using:

sudo pip3 install rpi_lcd

Now that we have these installed, it is time to pick an IDE (Integrated Development Environment). My personal favorite is Geany. You can install it from the drop-down menu.

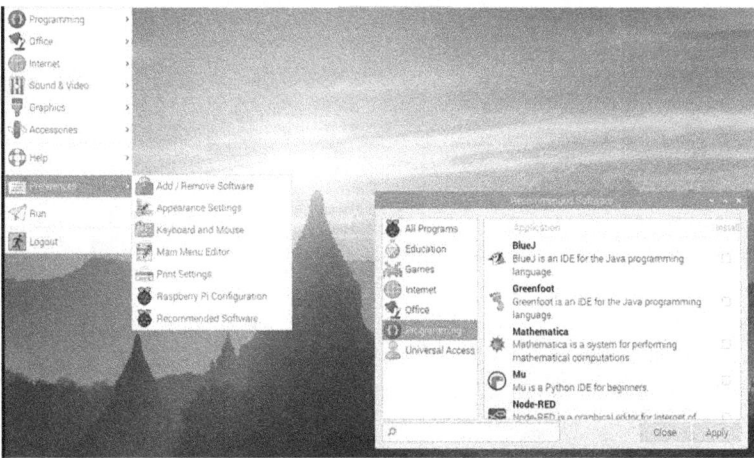

Figure 3-7. Installing Geany

I like Geany because once you save the file you are editing as a '.py' file it will help you with formatting such as indenting and different colors for different components of your program, such as variables and comments. It also does simple syntax checking. Before we can use Geany as an IDE, there is some setup we must do. Click on the "Build" tab on the task bar and select "Set Build Commands"

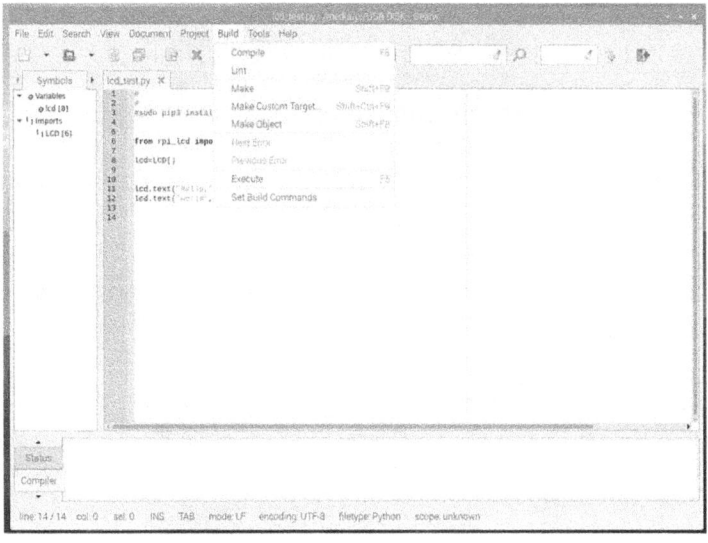

Figure 3-8. Set Build Commands

You will see the following pop-up window:

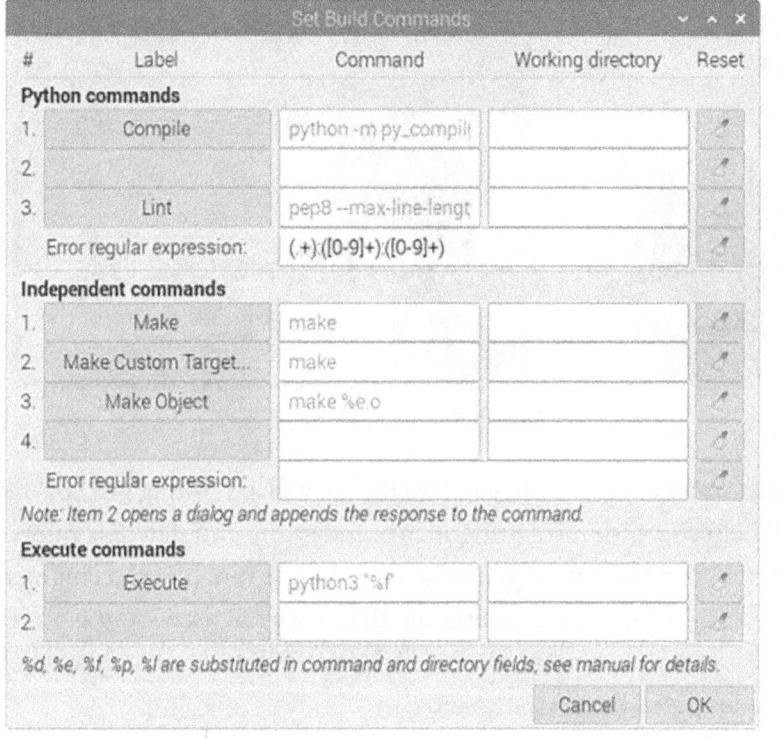

Figure 3-9. Pop Up Window

We need to enter the following information into the IDE.

Under the Python Commands section, line 1 (Compile) should read: python -m py_compile "%f"

Under the Execute commands section, line 1 (Execute) should read: python3 "%f"

From here on, I will assume that we are using Geany to write and run our programs.

Create a new file by clicking on the far-left icon on the task bar. This will open an empty file. At this point you should save the file as a python script such as "lcd_test.py". This will tell the IDE that you are writing a

Python program and help you with formatting and component colors. Enter the following code.

```
from rpi_lcd import LCD
lcd=LCD()
lcd.text("Hello",1)
lcd.text("World",2)
```

If you now click on the icon of the paper airplane, the program will run in the IDE environment.

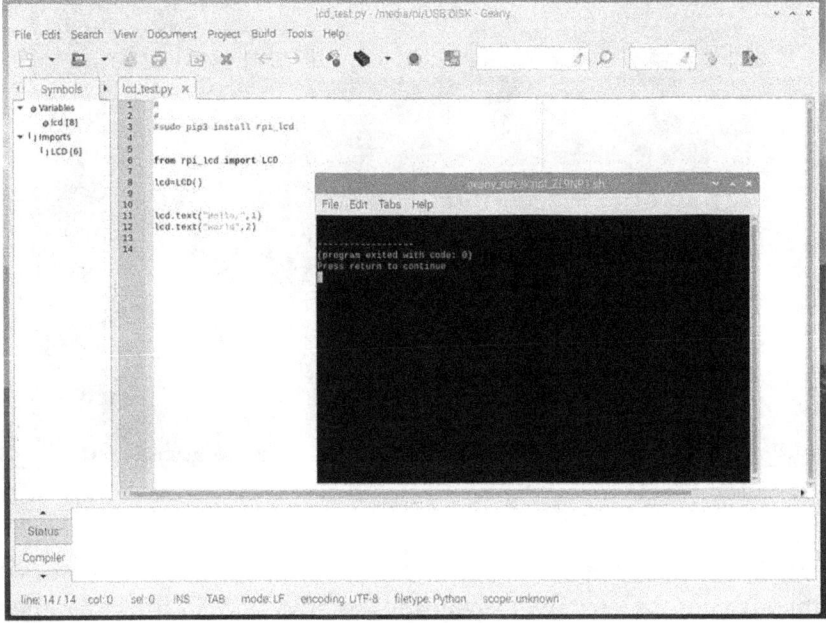

Figure 3-10. Running the test program

If there are no typing errors, the program will run, and you should see the following displayed on the LCD display:

Hello
world

If you see these words displayed on the LCD then the screen is connected properly, and everything is working properly. You may have to adjust the contrast, using the potentiometer on the back of the LCD.

If you don't see anything on the display and your electrical connections are correct, then it is trouble shooting time. The first thing to check is the address of the LCD on the I2C bus. We do this by opening a terminal window and entering the following command: i2cdetect -y 1

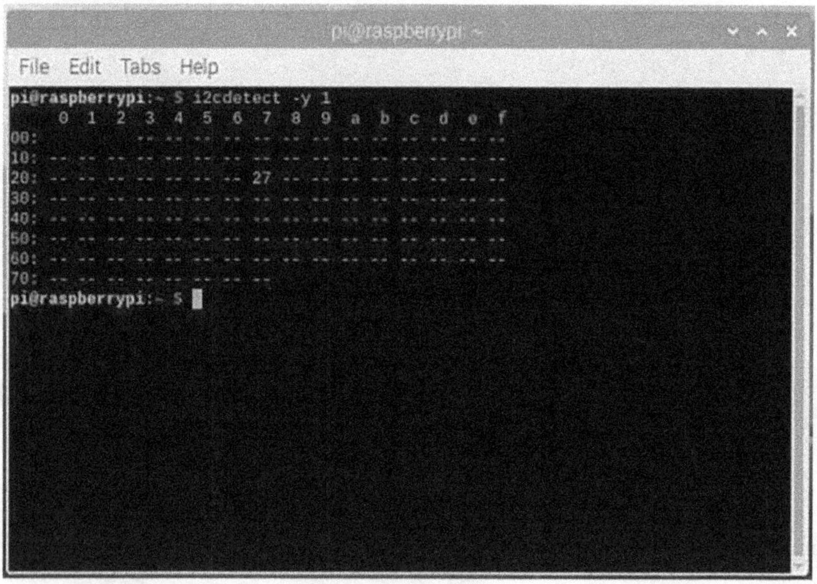

Figure 3-11. i2cdetect

If your connections are correct, the display should appear in the address table. If you don't see the display address anywhere in the table, then the display is not being detected for some reason and may be damaged.

The default address of the LCD display should be "0x27" hex. If you see a different address, then we will have to modify our code slightly.

The LCD object can be created with the address as parameter: lcd = LCD(address=0x3f) or what ever address is shown by i2cdetect.

Human Interface

Voice Output:

The alarm system software is capable of speaking to you in one of four languages; English, French, German or Dutch. I am an English speaker so I had to use an online translation program to get the other three languages. I aplolgise if the translation is not the best. The software package we will be using parses a text string and 'speaks' what ever is in the text string using a phoneme based library. That program is called 'espeak' and here is how we go about using it. The first thing we have to do is to install the espeak package using the following command:

> sudo apt-get install -y espeak

This version of espeak is the command line version. The Python3 version is broken in the Buster version of Raspian, so we will be using a work around. We will be piping the output of espeak to Alsa Audio so we will have to install that package as well:

> sudo apt-get install alsaaud*

I will be covering the basic commands used by the alarm system in this chapter. A more complete description of the commands and languages is available from: http://espeak.sourceforge.net The following program will allow the user to test the espeak software with various voices and languages.

```
#
#sudo apt-get install -y espeak
#
#This program is broken in the latest version of the OS (Buster)
#sudo apt-get install -y python3-espeak
#

import os
import time
os.popen('espeak -s100')
while True:

        print("English")
        speak = "Alarm detected in Zone 1" #English
        os.popen('espeak -ven+f5 "' + speak + '" --stdout | aplay 2> /dev/null').read()
        time.sleep(5)

        print("French")
        speak = "Alarme detectee dans la zone 1" #French
        os.popen('espeak -vfr+f5 "' + speak + '" --stdout | aplay 2> /dev/null').read()
        time.sleep(5)

        print("German")
        speak = "Alarm in zone eins entdeckt" #German
        os.popen('espeak -vde+f5 "' + speak + '" --stdout | aplay 2> /dev/null').read()
        time.sleep(5)

        print("Dutch")
        speak = "Alarm gedetecteerd in zone een" #Dutch
        os.popen('espeak -vnl+f5 "' + speak + '" --stdout | aplay 2> /dev/null').read()
        time.sleep(5)
```

The first thing we have to do is to import the 'os' package so that we can pipe commands to alsa.

Next we import time so that we can sleep for 5 seconds in between statements.

Espeak tends to talk rather fast so we slow it down with the command 'os.popen('espeak -s100')'

This sets the rate to 100 words per minute. You can play with this command to suit your language.

These lines cause espeak to actually "speak" the text string sent to it:

> speak = "Alarm detected in Zone 1" #English
> os.popen('espeak -ven+f5 "' + speak + '" --stdout | aplay 2> /dev/null').read()

The espeak command tells espeak to speak English with female voice 5. This is piped to the alsa application (aplay). There are several male and female voices available (+f1 thru +f5 and +m1 thru +m7). I found the male voices sounded too much like a robot. Feel free to test them all.

Human Interface:
RFID Card Reader:

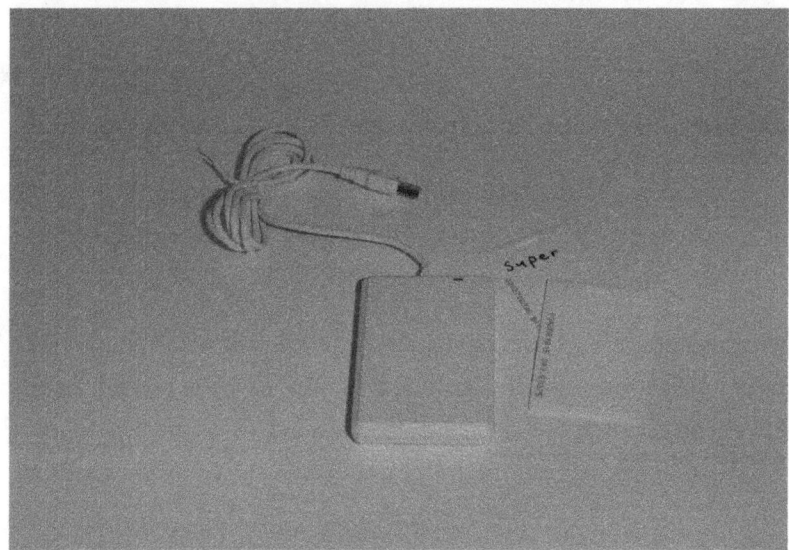

Figure 3-12. USB RFID Card Reader

This project uses an RFID card reader that plugs into a USB port on the Raspberry Pi. The reader appears to the operating system as a HID or Human Interface Device. Basically, the Raspberry Pi thinks that it is a keyboard. Before we can use the reader to arm and disarm the system, we must create a data base of known RFID cards. In our case it will be a simple text file. We will create this file using the text editor that comes with the system. To do this we select Accessories from the dropdown menu.

Figure 3-13. Select Text Editor

Next, we open a text file called "cards.txt". This will be our data base of acceptable cards. Adding cards to the database is a simple matter of swiping the cards across the card reader. The reader acts like a keyboard and sends a text string that corresponds to the serial number of the card, followed by a carriage return.

Figure 3-14. Enter cards in Database

Once you have all the cards entered into the file, save the file in the same directory as the Python script. Now it is time to create a special database file called "super.txt". This database will contain Supervisor cards. Only supervisor cards can arm the system. Once the system is armed, it can be disarmed by any card or key. Once the system is no longer in the 'System Idle" state it can be armed or disarmed using the key switch. Rotate the key to the left to disarm the system. Rotate the key to the right to arm the system. The key can only be removed if it is in the center position.

If you run the script the LCD display will say "System Idle". This means that the system is waiting to be armed. If you tap the card reader with a valid supervisor card, the LCD will display "System Armed" and the alarm system voice will also say "System Armed". The system is now capable of detecting intruders and fires. To disarm the system, tap a card again. The system will now say, and display "System Disarmed".

The fire detection system is different from the burglar alarm portion of the panel. The panel will still detect fires using the sensors connected to the fire zones even though the burglar alarm is disarmed. If a fire (heat rise) or smoke (smoke detector) is detected the alarm siren will sound. The siren can only be silenced by tapping the card reader twice with a supervisor card or turning the alarm key switch to the Disarm position.

Here is a typical scenario:

You disarm the burglar alarm because you are going to bed for the night, and you don't want someone to trigger the alarm because they got up for a glass of water. Unfortunately, someone has left something on in the kitchen, which starts a fire. Depending on the type of fire, either the heat rise (flame) detector or the smoke detector will trigger an alarm. You now have time to investigate the source of the alarm and to take appropriate action.

Chapter 4: Software

Version 6.0

In this chapter we will discuss the Python 3 software in detail, so that the builder can modify the code if necessary to suit their needs. This version uses Zone 1 as a special zone. It is intended that the user enters via a door whose contacts are connected to Zone 1 of the panel. The user has 2 minutes to reach the disarm key or the RFID reader to disarm the system. The system can be disarmed by either a supervisor card or a normal card.

The first few lines of code describe the modules which must be installed in order to access the Raspberry Pi hardware.

```
#
#sudo apt-get install alsaaud*
#sudo pip3 install rpi_lcd
#sudo apt-get install -y espeak
#
```

Next we define the registers used by the MCP23017 port expander IC.

```
# Define MCP23017 registers values from datasheet
IODIRA = 0x00 # IO direction A - 1= input 0 = output
IODIRB = 0x01 # IO direction B - 1= input 0 = output
IPOLA = 0x02 # Input polarity A
IPOLB = 0x03 # Input polarity B
GPINTENA = 0x04 # Interrupt-onchange A
GPINTENB = 0x05 # Interrupt-onchange B
DEFVALA = 0x06 # Default value for port A
DEFVALB = 0x07 # Default value for port B
```

```
INTCONA = 0x08 # Interrupt control register for port A
INTCONB = 0x09 # Interrupt control register for port B
IOCON = 0x0A # Configuration register
GPPUA = 0x0C # Pull-up resistors for port A
GPPUB = 0x0D # Pull-up resistors for port B
INTFA = 0x0E # Interrupt condition for port A
INTFB = 0x0F # Interrupt condition for port B
NTCAPA = 0x10 # Interrupt capture for port A
INTCAPB = 0x11 # Interrupt capture for port B
GPIOA = 0x12 # Data port A
GPIOB = 0x13 # Data port B
OLATA = 0x14 # Output latches A
OLATB = 0x15 # Output latches B
```

We then import the modules that will be used by the program.

```
import RPi.GPIO as GPIO
import time
import os
import threading
from rpi_lcd import LCD
from espeak import espeak
from smbus import SMBus
lcd=LCD()
```

We then setup the I2C bus for expander card 0.

```
# Setup I2C bus for Expander Card 0
i2cbus = SMBus(1) # Create a new I2C bus
i2caddress = 0x20 # Address of MCP23017 device
```

```
#i2cbus.write_byte_data(i2caddress, IOCON, 0x02) # Update configura-
tion register

i2cbus.write_word_data(i2caddress, IODIRA, 0xFF00) # Set Port A as
outputs and Port B as inputs

i2cbus.write_word_data(i2caddress, GPPUB, 0xFFFF) # Set Port B pull
up resistors on
```

We then assign alarm panel pins to the various zones. The number following the # is the corresponding pin on the 40 pin connector J15 on the alarm panel PCB.

```
#Broadcom numbering scheme
GPIO.setmode(GPIO.BCM)
GPIO.setwarnings(False)

# Alarm Panel Pin assignments
zone1 = 16        #J15-36
zone2 = 17        #J15-11
zone3 = 18        #J15-12
zone4 = 19        #J15-35
zone5 = 20        #J15-38
zone6 = 21        #J15-40
zone7 = 22        #J15-15
zone8 = 23        #J15-16
tamper = 25       #J15-22
arm = 24          #J15-18
disarm = 26       #J15-37
alarm = 27        #J15-13
fire1 = 5         #J15-29
fire2 = 6         #J15-31
```

Next we setup the Text To Speech strings. These are the words that will be spoken by the panel when a particular event occurs. These strings can be modified by the programmer to suit their needs. For example a different language.

```
# TTS Speech Strings
zone1_txt = "Motion detected in Zone 1"
zone2_txt = "Alarm detected in Zone 2"
zone3_txt = "Alarm detected in Zone 3"
zone4_txt = "Alarm detected in Zone 4"
zone5_txt = "Alarm detected in Zone 5"
zone6_txt = "Alarm detected in Zone 6"
zone7_txt = "Alarm detected in Zone 7"
zone8_txt = "Alarm detected in Zone 8"
arm_txt = "System Armed"
disarm_txt = "System Disarmed"
tamper_txt ="Tamper switch activated"
fire_txt = "Fire Detected"
```

We can now setup the function of the zone pins using GPIO.setup to setup the pins.

```
#Pin Setup
GPIO.setup(alarm,GPIO.OUT) #Set Alarm pin as an output

#Setup Zone pins as inputs with internal pull up
GPIO.setup(zone1, GPIO.IN, pull_up_down=GPIO.PUD_UP)
GPIO.setup(zone2, GPIO.IN, pull_up_down=GPIO.PUD_UP)
GPIO.setup(zone3, GPIO.IN, pull_up_down=GPIO.PUD_UP)
GPIO.setup(zone4, GPIO.IN, pull_up_down=GPIO.PUD_UP)
GPIO.setup(zone5, GPIO.IN, pull_up_down=GPIO.PUD_UP)
GPIO.setup(zone6, GPIO.IN, pull_up_down=GPIO.PUD_UP)
```

```
GPIO.setup(zone7, GPIO.IN, pull_up_down=GPIO.PUD_UP)
GPIO.setup(zone8, GPIO.IN, pull_up_down=GPIO.PUD_UP)
GPIO.setup(arm, GPIO.IN, pull_up_down=GPIO.PUD_UP)
GPIO.setup(disarm, GPIO.IN, pull_up_down=GPIO.PUD_UP)
GPIO.setup(tamper, GPIO.IN, pull_up_down=GPIO.PUD_UP)
GPIO.setup(fire1, GPIO.IN, pull_up_down=GPIO.PUD_UP)
GPIO.setup(fire2, GPIO.IN, pull_up_down=GPIO.PUD_UP)
```

Now we can configure the alarm system for its idle state.

```
GPIO.output(alarm,GPIO.LOW)
lcd.clear
lcd.text(" System Idle",1)
arm_flg = False
Card_flg = False
espeak.synth("Alarm System Idle")
```

There is currently a problem with the espeak software under the current revision of the Linux operating system. Espeak sometimes does not 'speak' the first word of the phrase it is given. It works fine after the first sentence. For that reason, the above command was added at the end of the setup sequence.

The following functions are used to perform various tasks. The first function interfaces with the RFID card reader.

```
def CardRead():
 global Card_flg
 global card_id
 card_id = input()
 Card_flg= True
 threading.Timer(1, CardRead).start()
```

This function checks to see if someone has entered through the main door of the house, office or lab. It first checks to see if the system has been disarmed with an alarm system key. It then checks to see if the door has been opened. If the zone 1 input is high (1), it exits the function. If not, the input must be low signaling that the door is open. It informs the user that they have two minutes to disarm the system. It does this using the LCD display and Espeak to announce the 'zone1_txt' message. It does this using an English, female voice. This can be changed by the programmer to speak several different languages.

```
def Entry():
        global arm_flg
#Check Disarm Switch
        if GPIO.input(disarm):
                time.sleep(.1)
        else:
                lcd.clear()
                lcd.text("System Disarmed",1)
                arm_flg=0
                GPIO.output(alarm,GPIO.LOW) #Turn off siren
```

```
#Check Zone 1

        if GPIO.input(zone1):

                pass

        else:

                lcd.text("Motion Detected",1)

                lcd.text("You have 2 min",2)

                os.popen('espeak -ven+f5 "' + zone1_txt + '" --std
                out | aplay 2> /dev/null').read()

                time.sleep(120)

                lcd.text("Your 2 min is up",2)

                threading.Timer(1, Entry).start()
```

This function checks the fire zone inputs. These are special zones which override other functions of the code. The fire function will signal a fire even if the system is disarmed.

```
def fire():

#Check Fire Zone 1

        if GPIO.input(fire1):

                pass

        else:

                lcd.text(" FIRE !",1)

                lcd.text("Fire Zone 1",2)

                os.popen('espeak -ven+f5 "' + fire_txt + '" --stdout
                | aplay 2> /dev/null').read()

                arm_flg = True

                GPIO.output(alarm,GPIO.HIGH) #Turn on Siren relay

#Check Fire Zone 2

        if GPIO.input(fire2):

                pass

        else:
```

```
lcd.text(" FIRE !",1)

lcd.text("Fire Zone 2",2)

os.popen('espeak -ven+f5 "' + fire_txt + '" --stdout
| aplay 2> /dev/null').read()

arm_flg = True

GPIO.output(alarm,GPIO.HIGH) #Turn on Siren relay
threading.Timer(2, fire).start()
```

This function is used to read from an expander card if one is attached to the system. This version of the software only checks for one expander card. Up to five other cards can be accessed by modifying the code shown below. The first few lines of code read from the port expander IC and checks bit 0. This code assumes that a light/darkness detector is connected to port 0 of the expander board. If the pin is low the code turns on output one which is connected to an LED flood light. If the input is high, then it assumes that light has been detected and turns the flood light off. This is a simple example of how to control external devices using the expander board. The other inputs are read in a similar fashion and a test message is sent to the console for testing purposes.

```
def Expander1():
        portb = i2cbus.read_byte_data(i2caddress, GPIOB) # Read the
        value of Port B
        portb_bin = (int(portb))
        # See if a bit 0 is set
        bit_mask = int('0b00000001', 2) # Bit 0
        # Is bit set in byte1?
        result = (bit_mask & portb_bin)
        if result :
                i2cbus.write_byte_data(i2caddress, GPIOA,
                0x00) # Set pin 1 to off
                pass

        else:
                print("Darkness Detected")
                i2cbus.write_byte_data(i2caddress, GPIOA,
                0x01) # Set pin 1 to on
```

```
# See if a bit 1 is set
        bit_mask = int('0b00000010', 2) # Bit 1
        # Is bit set in byte1?
        result = (bit_mask & portb_bin)
        if result :
                pass
        else:
                print("Alarm in Zone 10")

# See if a bit 2 is set
        bit_mask = int('0b00000100', 2) # Bit 2
        # Is bit set in byte1?
        result = (bit_mask & portb_bin)
        if result :
                pass
        else:
                print("Alarm in Zone 11")

# See if a bit 3 is set
        bit_mask = int('0b00001000', 2) # Bit 3
        # Is bit set in byte1?
        result = (bit_mask & portb_bin)
        if result :
                pass
        else:
                print("Alarm in Zone 12")

# See if a bit 4 is set
        bit_mask = int('0b00010000', 2) # Bit 4
        # Is bit set in byte1?
        result = (bit_mask & portb_bin)
        if result :
                pass
        else:
                print("Alarm in Zone 13")

# See if a bit 5 is set
        bit_mask = int('0b00100000', 2) # Bit 5
        # Is bit set in byte1?
```

```
                result = (bit_mask & portb_bin)
                if result :
                        pass
                else:
                        print("Alarm in Zone 14")

        # See if a bit 6 is set
                bit_mask = int('0b01000000', 2) # Bit 6
                # Is bit set in byte1?
                result = (bit_mask & portb_bin)
                if result :
                        pass
                else:
                        print("Alarm in Zone 15")

        # See if a bit 7 is set
                bit_mask = int('0b10000000', 2) # Bit 7
                # Is bit set in byte1?
                result = (bit_mask & portb_bin)
                if result :
                        pass
                else:
                        print("Alarm in Zone 16")
```

The following code calls the functions we defined earlier.

```
CardRead()       #Comment out this line to use arm switch ONLY
fire()           #Check for fires
Entry()          #Check Zones 1 to 8 and Disarm Key
```

If the system is not armed keep checking the alarm key switch and the card reader to see if it is, by first checking the alarm key switch.

```
while arm_flg == False:
        #If we are not armed, check Arm switch
        if GPIO.input(arm):
                pass
        else:
                lcd.clear()
                lcd.text(" System Armed",1)
```

```
lcd.text(" In 2 Minutes",2)
time.sleep(120)
lcd.text(" ",2)
os.popen('espeak -ven+f5 "' + arm_txt + '" --stdout
| aplay 2> /dev/null').read()
arm_flg=True
```

Next, we check to see if a card has been swiped.

```
#See if a supervisor card has been swiped
        if Card_flg == True:
                id_file = open("super.txt","r")
                for line in id_file:

                        if (card_id == line.strip()):
                                lcd.clear()
                                lcd.text(" System Armed",1)
                                lcd.text(" In 2 Minutes",2)
                                time.sleep(120)
                                lcd.text(" ",2)
                                os.popen('espeak -ven+f5 "' + arm_txt + '"
                                -stdout | aplay 2> /dev/null').read()
                                arm_flg = True
                                Card_flg = False
                        else:
                                pass
```

Once the system is armed with either an alarm key or a supervisor card, we enter the main loop.

```
while True:
        #If we are not armed, check Arm switch
        if GPIO.input(arm):
                time.sleep(.1)
        else:
                lcd.clear()
                lcd.text(" System Armed",1)
                lcd.text(" In 2 Minutes",2)
                time.sleep(120)
                lcd.text(" ",2)
                os.popen('espeak -ven+f5 "' + arm_txt + '" --stdout
                | aplay 2> /dev/null').read()
```

```python
                arm_flg=True
    if Card_flg == 1:
            id_file = open("super.txt","r")
            for line in id_file:
            if (card_id == line.strip()):
                    lcd.clear()
                    lcd.text(" System Armed",1)
                    lcd.text(" In 2 Minutes",2)
                    time.sleep(120)
                    lcd.text(" ",2)
                    os.popen('espeak -ven+f5 "' + arm_txt + '"
                    --stdout | aplay 2> /dev/null').read()
                            arm_flg = True
                            Card_flg = False
            else:
                    pass
#If we are armed, check zones
        while arm_flg == True:
#Check Zone 2
        if GPIO.input(zone2):
            pass
        else:
            lcd.text(" ALARM",1)
            lcd.text("Zone 2",2)
            os.popen('espeak -ven+f5 "' + zone2_txt + '" --std
            out | aplay 2> /dev/null').read()
            GPIO.output(alarm,GPIO.HIGH) #Turn on Siren relay
#Check Zone 3
        if GPIO.input(zone3):
            pass
        else:
            lcd.text(" ALARM",1)
            lcd.text("Zone 3",2)
            os.popen('espeak -ven+f5 "' + zone3_txt + '" --std
            out | aplay 2> /dev/null').read()
            GPIO.output(alarm,GPIO.HIGH) #Turn on Siren relay
#Check Zone 4
        if GPIO.input(zone4):
            pass
```

```
        else:
                lcd.text(" ALARM",1)
                lcd.text("Zone 4",2)
                os.popen('espeak -ven+f5 "' + zone4_txt + '" --std
                out | aplay 2> /dev/null').read()
                GPIO.output(alarm,GPIO.HIGH) #Turn on Siren relay
#Check Zone 5
        if GPIO.input(zone5):
                pass
        else:
                lcd.text(" ALARM",1)
                lcd.text("Zone 5",2)
                os.popen('espeak -ven+f5 "' + zone5_txt + '" --std
                out | aplay 2> /dev/null').read()
                GPIO.output(alarm,GPIO.HIGH) #Turn on Siren relay

#Check Zone 6
        if GPIO.input(zone6):
                pass
        else:
                lcd.text(" ALARM",1)
                lcd.text("Zone 6",2)
                os.popen('espeak -ven+f5 "' + zone6_txt + '" --std
                out | aplay 2> /dev/null').read()
                GPIO.output(alarm,GPIO.HIGH) #Turn on Siren relay
#Check Zone 7
        if GPIO.input(zone7):
                pass
        else:
                lcd.text(" ALARM",1)
                lcd.text("Zone 7",2)
                os.popen('espeak -ven+f5 "' + zone7_txt + '" --std
                out | aplay 2> /dev/null').read()
                GPIO.output(alarm,GPIO.HIGH) #Turn on Siren relay
#Check Zone 8
        if GPIO.input(zone8):
                pass
        else:
                lcd.text(" ALARM",1)
                lcd.text("Zone 8",2)
                os.popen('espeak -ven+f5 "' + zone8_txt + '" --std
```

```
              out | aplay 2> /dev/null').read()
              GPIO.output(alarm,GPIO.HIGH) #Turn on Siren relay
```

Once we have checked all of the zone inputs on the main board, we check the tamper switch, the alarm switch and the card reader.

```
#Check Tamper Switch
        if GPIO.input(tamper):
              lcd.text(" ALARM",1)

              lcd.text("Tamper Switch",2)

              os.popen('espeak -ven+f5 "' + tamper_txt + '"
              --stdout | aplay 2> /dev/null').read()

              GPIO.output(alarm,GPIO.HIGH) #Turn on Siren relay

#Check Disarm Switch
        if GPIO.input(disarm):
              time.sleep(.1)

        else:

              lcd.clear()

              lcd.text("System Disarmed",1)

              arm_flg=0

              GPIO.output(alarm,GPIO.LOW) #Turn off siren

              os.popen('espeak -ven+f5 "' + disarm_txt + '"
              --stdout | aplay 2> /dev/null').read()

#Check Arm Switch
        if GPIO.input(arm):
              time.sleep(.1)

        else:

              lcd.clear()

              lcd.text(" System Armed",1)

              lcd.text("In 2 Minutes",2)

              time.sleep(120)

              lcd.text(" ",2)
```

```
            os.popen('espeak -ven+f5 "' + arm_txt + '" --stdout
            | aplay 2> /dev/null').read()

            arm_flg= True

#Check Card Reader

        if Card_flg == True:

            id_file = open("cards.txt","r")

            for line in id_file:

                if (card_id == line.strip()):

                    lcd.clear()

                    lcd.text("System Disarmed",1)

                    os.popen('espeak -ven+f5 "' + disarm_txt +
                    '" --stdout | aplay 2> /dev/null').read()

                    GPIO.output(alarm,GPIO.LOW) #Turn off siren

                    arm_flg = False

                    Card_flg = False

        else:

            pass
```

Finally, we call the Expander function(s) to check the inputs on these boards.

```
# Check Expander 1 input ports

        Expander1()    #Check for expander card

        #Comment out this line if the is no expander card(s)
        attached
```

Speech_test.py

Also included with the code is a test program called speech_test.py so that the programmer can experiment with various voices, languages and speed.

```python
#https://espeak.sourceforge.net/index.html
#sudo apt-get install -y espeak
#
import os
import time
os.popen('espeak -s35') #This is the speech speed 35 words per minute.
while True:

        print("English")
        speak = "Alarm detected in Zone 1" #English
        os.popen('espeak -ven+f5 "' + speak + '" --stdout | aplay 2> /dev/null').read()
        time.sleep(5)


        print("French")
        speak = "Alarme detectee dans la zone 1" #French
        os.popen('espeak -vfr+f5 "' + speak + '" --stdout | aplay 2> /dev/null').read()
        time.sleep(5)


        print("Spanish")
        speak = "Alarma detectada en Zona 1" #Spanish
        os.popen('espeak -ves+f5 "' + speak + '" --stdout | aplay 2> /dev/null').read()
        time.sleep(5)
        print("German")
        speak = "Alarm in zone eins entdeckt" #German
```

```
os.popen('espeak -vde+f5 "' + speak + '" --stdout | aplay 2> /dev/null').read()
time.sleep(5)


print("Dutch")
speak = "Alarm gedetecteerd in zone een" #Dutch
os.popen('espeak -vnl+f5 "' + speak + '" --stdout | aplay 2> /dev/null').read()
time.sleep(5)
```

TestPorts.py

This diagnostic program can be used by the reader to turn the outputs of the alarm system on and off. It also serves as a programming example for turning an output on or off in the main code.

```
#
#This program turns the alarm system outputs on and off
#
import RPi.GPIO as GPIO

GPIO.setmode(GPIO.BCM)

GPIO.setwarnings(False)

GPIO.setup(27,GPIO.OUT) #Set alarm pin as an output

# Define registers values from datasheet

IODIRA = 0x00 # IO direction A - 1= input 0 = output

IODIRB = 0x01 # IO direction B - 1= input 0 = output

IPOLA = 0x02 # Input polarity A

IPOLB = 0x03 # Input polarity B

GPINTENA = 0x04 # Interrupt-onchange A

GPINTENB = 0x05 # Interrupt-onchange B

DEFVALA = 0x06 # Default value for port A

DEFVALB = 0x07 # Default value for port B

INTCONA = 0x08 # Interrupt control register for port A

INTCONB = 0x09 # Interrupt control register for port B

IOCON = 0x0A # Configuration register

GPPUA = 0x0C # Pull-up resistors for port A

GPPUB = 0x0D # Pull-up resistors for port B

INTFA = 0x0E # Interrupt condition for port A

INTFB = 0x0F # Interrupt condition for port B

INTCAPA = 0x10 # Interrupt capture for port A
```

```
INTCAPB = 0x11 # Interrupt capture for port B

GPIOA = 0x12 # Data port A

GPIOB = 0x13 # Data port B

OLATA = 0x14 # Output latches A

OLATB = 0x15 # Output latches B


from smbus import SMBus

import time


i2cbus = SMBus(1) # Create a new I2C bus

i2caddress = 0x20 # Address of first MCP23017 device


i2cbus.write_word_data(i2caddress, IODIRA, 0xFF00) # Set Port A as outputs and Port B as inputs

i2cbus.write_word_data(i2caddress, GPPUB, 0xFFFF) # Set Port B pull up resistors on


while(True):

    print ("Do you want to turn on the siren? ")

    siren = input("[Yes] or [No] ")

    if siren == "yes":

        GPIO.output(27,GPIO.HIGH) #Turn siren on

        pass

    if siren == "no":

        GPIO.output(27,GPIO.LOW) #Turn siren off

        Pass

    port = int(input("Enter an output between 0 and 7: "))

    if port >= 0:

        if port < 8:

            status = int(input("Enter 1 to turn on the output or 0 to turn off."))
```

```
else: print ("Invalid port number")

pass

if port == 0:

# Bit 0

if status == 1:

i2cbus.write_byte_data(i2caddress, GPIOA, 0x01) # Set PA0 to on

pass

if status == 0:

i2cbus.write_byte_data(i2caddress, GPIOA, 0x00) # Set PA0 to off

pass

#Bit 1

if port == 1:

if status == 1:

i2cbus.write_byte_data(i2caddress, GPIOA, 0x02) # Set PA1 to on

pass

if status == 0:

i2cbus.write_byte_data(i2caddress, GPIOA, 0x00) # Set PA1 to off

pass

#Bit 2

if port == 2:

if status == 1:

i2cbus.write_byte_data(i2caddress, GPIOA, 0x04) # Set PA2 to on

pass

if status == 0:

i2cbus.write_byte_data(i2caddress, GPIOA, 0x00) # Set PA2 to off

pass

#Bit 3

if port == 3:
```

```
if status == 1:
    i2cbus.write_byte_data(i2caddress, GPIOA, 0x08) # Set PA3 to on
    pass
if status == 0:
    i2cbus.write_byte_data(i2caddress, GPIOA, 0x00) # Set PA3 to off
    pass
#Bit 4
if port == 4:
    if status == 1:
        i2cbus.write_byte_data(i2caddress, GPIOA, 0x10) # Set PA4 to on
        pass
    if status == 0:
        i2cbus.write_byte_data(i2caddress, GPIOA, 0x00) # Set PA4 to off
        pass
#Bit 5
if port == 5:
    if status == 1:
        i2cbus.write_byte_data(i2caddress, GPIOA, 0x20) # Set PA5 to on
        pass
    if status == 0:
        i2cbus.write_byte_data(i2caddress, GPIOA, 0x00) # Set PA5 to off
        pass
#Bit 6
if port == 6:
    if status == 1:
        i2cbus.write_byte_data(i2caddress, GPIOA, 0x40) # Set PA6 to on
        pass
    if status == 0:
```

```
i2cbus.write_byte_data(i2caddress, GPIOA, 0x00) # Set PA6 to off

pass

#Bit 7

if port == 7:

if status == 1:

i2cbus.write_byte_data(i2caddress, GPIOA, 0x80) # Set PA7 to on

pass

if status == 0:

i2cbus.write_byte_data(i2caddress, GPIOA, 0x00) # Set PA7 to off
```

Chapter 5: Printed Circuit Board

Assembly

In this chapter we will cover assembly and test of the printed circuit board. You will require a fine tipped soldering iron, solder, and a sharp pair of diagonal cutters. You may also find that a solder sucker and solder wick come in handy. The figure below is the silk screen layer of the circuit board. You should refer to the bill of materials for component types and values.

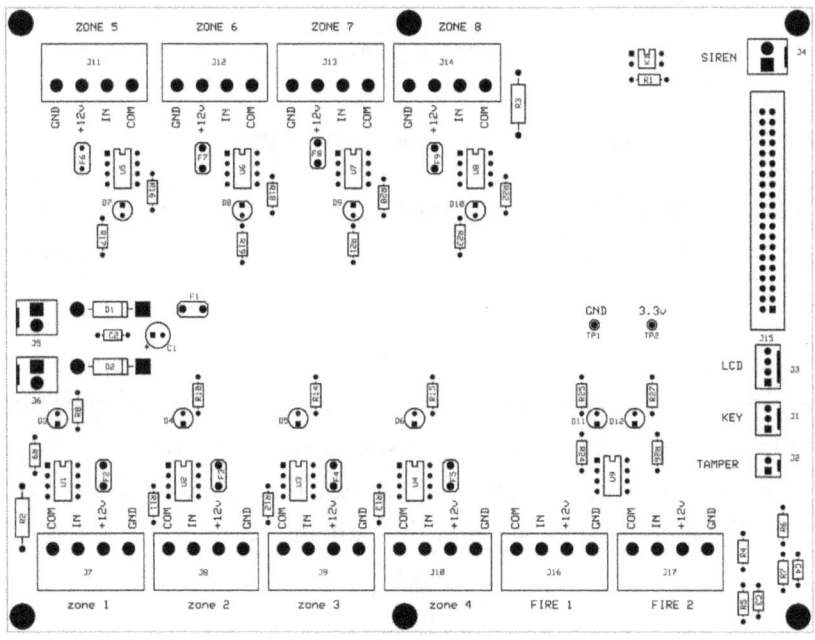

Figure 5-1. Silk Screen

We will start the assembly with the lowest (physically) components. In this case the ¼ watt resistors. There are 27 of them so it will take some time to solder them all in place. I recommend that you solder them one at a time, by inserting the resistor on the silk screen side and then flipping the board over to solder the component in place. The board has solder mask on both sides, but care must be taken so as not to short the component pins to the ground plane.

The next components we will be soldering in place is the ICs. You can use the same procedure as the resistors. Note that pin 1 of the IC is a square pad on the printed circuit board. The silk screen has a notch drawn to indicate pin 1. The IC package itself should have a dot indicating pin 1. A magnifying glass may be useful to read the pin 1 marker and the part name.

Figure 5-2. Integrated Circuit Markings

I recommend soldering diagonal corners, checking that the IC is seated properly and then soldering the other pins. In this case pin four should be soldered to the ground plane.

The next component to install is the red LEDs. These components can be tricky to install because the way the polarity is indicated varies from one manufacturer to another. Some use a flat side to indicate the cathode, and some don't. In general, the shortest lead is the cathode. In each case the LED should be installed with the cathode facing the IC.

Figure 5-3. LED

Case and lead styles vary from one manufacturer to another. Depending on where you purchased your parts, the LED may not fit flush with the PCB. This is not important, so long as it is soldered in place properly.

An LED is basically a diode which emits light when it is forward biased. That is why it has a cathode and an anode. If you are not sure of the polarity of your devices, they can be checked like any other diode. Using the following method.

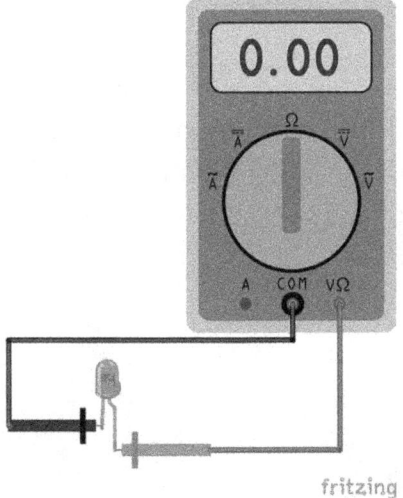

Figure 5-4. Testing an LED

Set the digital multimeter to the lowest ohms scale available. Connect the black lead of the DMM to what you think is the cathode. Connect the red lead to the anode lead. The LED should glow dimly, and you should see a very low reading on the DMM. An LED is like any other diode when it comes to DMMs. If you get a high reading, reverse the leads.

The next component to install is the PTC fuses. These are not fuses in the normal sense. They do not 'blow' and have to be replaced like normal fuses you would find in other types of equipment or devices.

The letters 'PTC' stands for Positive Temperature Coefficient. This means that a PTC fuse acts like a variable resistor. The more current it passes the hotter it gets and the higher the resistance becomes. This continues until the avalanche point, at which point it acts like normal fuse and the resistance increases dramatically. After a certain period of time, if the high current is removed, the device cools down and 'heals' itself. That is why they are being used in this alarm system. That way there is no need to open the panel to replace conventional fuses.

These types of fuses are thermal devices so care should be taken when soldering them, so as not to damage the component. Use a fine soldering tip as possible.

The next component to install is J15. This is the 40-pin ribbon cable connector which connects the panel motherboard to the Raspberry Pi. Pin 1 is marked with an arrow and there is a polarizing notch for the ribbon cable connector. Pin 1 on the PCB is marked with a square pad.

Figure 5-5. 40 Pin Ribbon Connector

Care must be taken when soldering this component. The pins are on .100" spacing, however some have tracks running between them. A visual inspection with a magnifying glass, after soldering could very well save you the price of a Raspberry Pi.

Connectors J1, J2 and J3 can now be installed. These are the connections to the Key switch, the Tamper switch, and the LCD display. The connectors should be installed with the tab facing the edge of the board. As before, I suggest soldering one pin and then checking that the connector is seated properly before soldering the remaining pins.

Components C1, C2, D1 and D2 can now be installed as well. Note that the cathodes of D1 and D2 are marked with a stripe and that the cathode pin is square on the PCB. The other thing to be aware of is the polarity of capacitor C1. The positive terminal is marked on the PCB and is a square pad on the PCB.

Connectors J4, J5 and J6 can now be installed as well. Once again, the tabs should be facing the edge of the PCB.

At this point you may want to change to a medium sized tip on your soldering iron. This is because the pins on the screw terminal blocks are rather large, compared to the pins we have been working with so far.

Figure 5-6. Four terminal block

When installing the terminal blocks, note that the open side of the block should face the edge of the PCB.

We have one last temporary component to install and that is the AA cell holder. If you purchased the holder listed in the Bill of Materials, it will have two red and black wires attached to it. Solder the red wire to TP2-3.3v and the black wire to TP1-GND. If you like you can fasten the holder to the PCB with a small piece of double-sided tape.

Figure 5-7. Battery Holder

Printed Circuit Board

Testing

Now that we have the PCB assembled, it is time to begin testing. To complete the tests, I recommend the following tools. A Digital Volt Meter (DVM), needle nose plyers, and a 12 volt power supply of some sort. The power supply should be a current limited bench power supply and a couple of miniature test leads. If you do not have access to a bench supply, then a modified 12-volt adapter will do. Simply cut the DC power connector from the adapter and solder a couple of clip leads to the adapter.

Figure 5-8. Clip Leads

Once this is done, it is time to start powering up the board. First, we will install the alkaline batteries. Note that it is VERY important that you use only alkaline cells. This is because they will survive a short circuit, while other chemistries like lithium will overheat and blow up!

Measure the voltage between TP1 and TP2. You should read approximately 3 volts DC. All that the 3V is doing is powering the LED's so the actual voltage is not critical. Just make sure that the batteries are installed correctly, and that 3V appears on TP2.

It is now time to power up the alarm panel. Connect the red (positive) clip lead from your power supply to pin 1 of either J5 or J6. Connect the black (negative) clip lead to pin 2 of either J5 or J6. Note that pin one of the connectors is connected to the anode of D1 or D2.

Plug in or turn on your 12V power supply, and immediately check the voltage between the cathodes of D1 and D2 and ground. You should read about 12VDC if you are using an adapter or whatever you have your bench supply set to. (12VDC @ 1A should do.) If you do not read about 12VDC, quickly remove the power, and begin the trouble shooting portion of this chapter.

There is very little that can go wrong at this point. Check the polarity of C1. It is an electrolytic capacitor, and the positive terminal should be connected the cathodes of D1 and D2. The only other possibility is a short circuit. You may need some insulating tape on the clip leads to prevent short circuits. We are not talking lethal voltages here, so any kind of insulating tape will do. You may want to consider purchasing some nylon stand offs or rubber feet for the bottom of the PCB, to prevent short circuits to your work bench. I prefer a wooden work bench for just this reason.

You should see eight LEDs lit when you apply the 12-volt power. One for each alarm zone. This is because we do not have an end of line resistor installed yet, so the circuit thinks that an alarm contact is open. If you don't see all 8 LEDs lit, don't worry we are about to make some measurements to verify that all is well. We have already checked the input voltage at the cathode of the two power diodes, so now it is time to check the voltage on the other side of F1. We do this by measuring the voltage between the output side of F1 and ground. If you read the same voltage as the input, then all is well. If you measure considerably less than the input voltage, then something is drawing too much current and the fuse is doing its job. If you have a short circuit or an IC in backwards, the fuse may be warm to the touch. You can check the ICs with your finger to see if they are warm to the touch. Any way, it is time to remove the power and check for shorts and reversed ICs.

If the LEDs are lit and the 12-volt rail appears to be good, check pin 8 of all the ICs for 12 volts with respect to ground. If not, then you may have a bad solder joint. Now it is time to verify the reference voltages supplied to the comparators, starting with the 9-volt reference. This is produced by voltage divider R4 and R5.

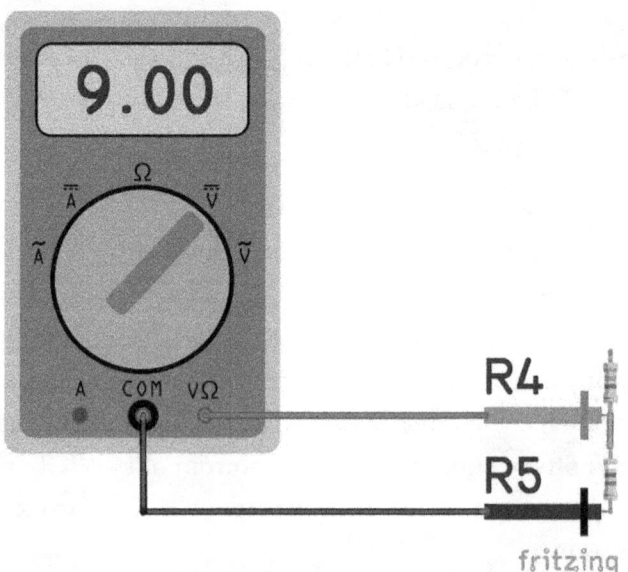

Figure 5-9. 9-volt reference

If the reference voltages are incorrect, then this may be a cause of no lit LEDs. You should measure 12 volts on the other side of R4. Check the resistor color codes. They should match the ones in the picture.

Now that we have checked the 9-volt reference, it is time to check the 2-volt reference. This reference is created by resistors R6 and R7. You should measure about 2 volts across R7.

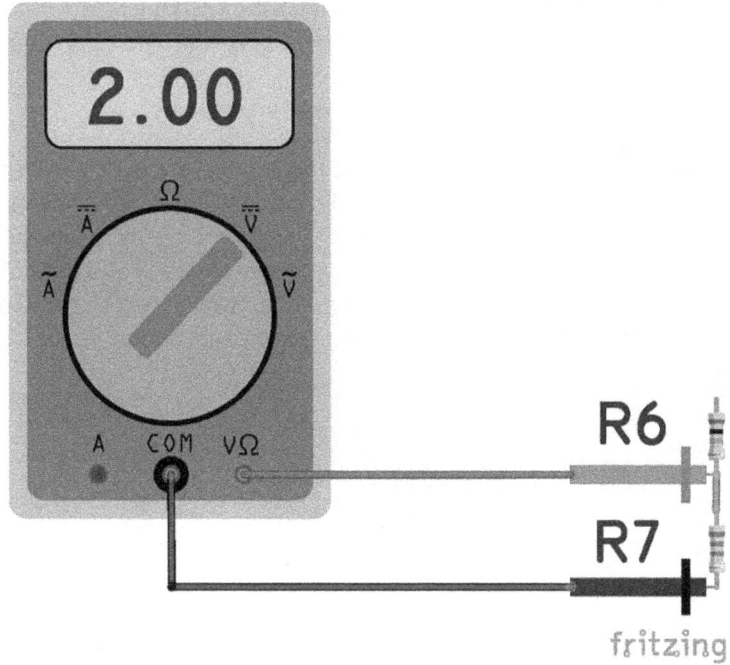

Figure 5-10. 2-volt Reference

Again, there should be 12 volts at the other end of R6. The reason why I have not given you absolute values for your measurements is that there are several tolerances at work here. First, unless you are using a bench supply, your 12-volt power will not be exactly 12 volts. Also, we are using 5% tolerance resistors, so even with a bench supply, your measurements could be off by as much as 10%. I have allowed for this in the design of the circuit. Capacitors C3 and C4 are not shown because they are decoupling capacitors and are there to reduce ripple and noise on the references. They do not affect the DC voltage.

The other two zones are 'Fire' zones and use U9. There should be a 9-volt reference on pins 3 and pin 5 of the IC. You should also see 12 volts on pin 8 of the IC. That is all the measurements we can do on the fire zones for now.

We can however start our zone testing by shorting the 'COM' pin of each zone to the 'IN' pin of the same zone. This applies 12 volts to the input of the comparator and should cause the LED to light. If it doesn't, make sure the polarity of the LED is correct. The cathode side of the LED should face the IC. Fire zones use normally open alarm contacts that close when an alarm condition is detected. There is no end of line resistor, so you can short the two terminals with a test lead.

There are several ways that we can test the other eight burglar alarm zone inputs. The easiest way is to first disconnect the 12-volt power, and then install eight 5K6 resistors between the 'COM' and 'IN' terminals of each zone. These resistors will eventually become our end of line resistors. A better way of testing the zones is to make up a tester like the one shown. The push button is normally closed and simulates the alarm contacts.

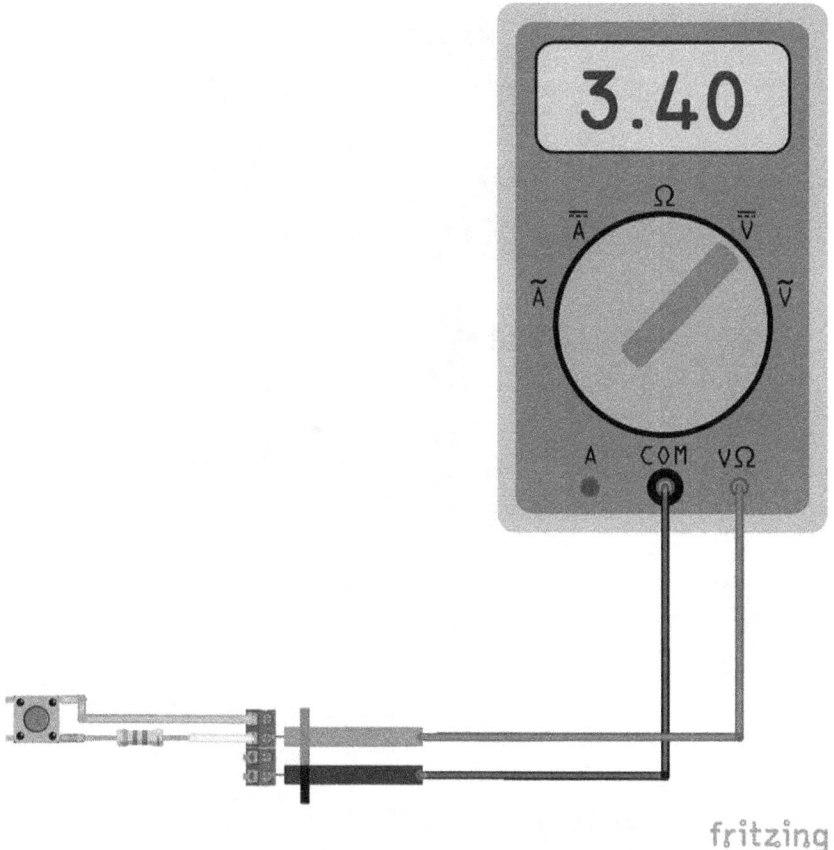

Figure 5-11. Testing Zone Inputs

With the switch closed (not pressed), you should read about 3.4 volts on the DMM, and the LED should be out. When you press the switch to open the contacts, you should read 0 volts, and the LED should light.

The final test is the 12-volt supply to the sensors. The burglar alarm zones have a PTC fuse protecting the 12-volt supply. The fire zones are protected by the main fuse.

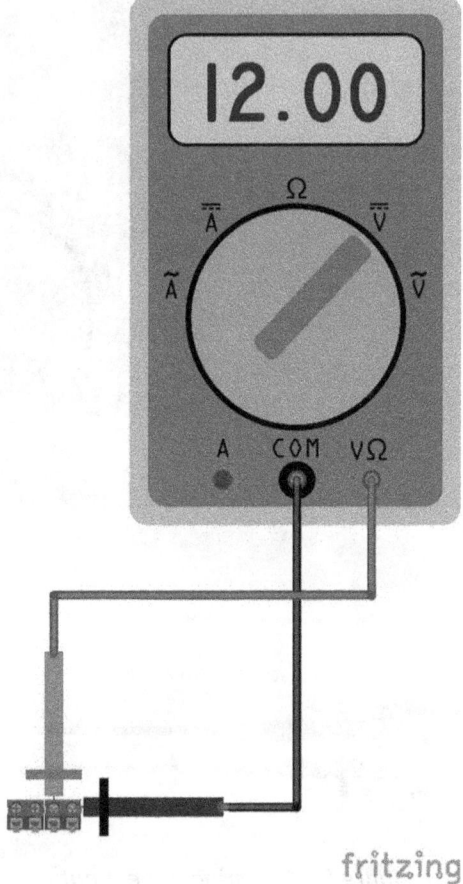

Figure 5-12. Testing 12-volt supply

In either case, you should measure about 12 volts between the terminal marked '+12v' and the ground terminal.

Chapter 6: Alarm System Wiring

Zone Test Switches

In this chapter you will be provided with tips on how to wire the various system sensors to the alarm panel PCB. If you have been following the text so far, then your PCB will look much like figure 1 below.

Figure 6-1. Alarm Panel PCB

You will notice that some of the zones have an end of line resistor connected between the Input and Common terminals. This is to prevent false alarms during testing of the software by making the alarm panel think that there is a sensor connected to the zone. The remaining zones will be connected to push button switches. These switches are used to simulate alarm system sensors for test purposes. The battery pack is used to supply 3-volt power for the LEDs.

The next thing we add is the zone alarm switches. The switches are connected between the Common and the Input terminals of each zone.

Figure 6-2. Zone Switches

Two types of switches are used to simulate alarm and fire sensors. The black buttons are normally open switches. When the button is pressed, it closes the circuit to simulate a fire detection.

The red buttons are normally closed and open when the button is pressed to simulate an alarm condition. These buttons are connected in series with an end of line resistor. One switch is connected to Zone 1. You may recall that this is a special zone which is used by the software to detect entry thru the outside door. The other switch can be moved from zone to zone to test the remaining zones. The LED associated with the zone will light when the button is pressed. All the other zones should have end of line resistors installed and their corresponding LEDs should be out.

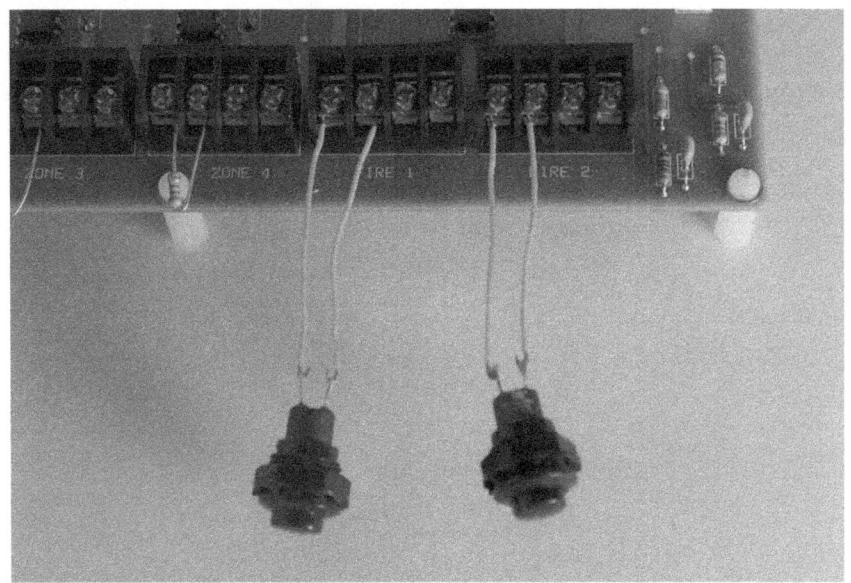

Figure 6-3. Normally Open Fire Zones

Figure 6-4. Normally Closed Alarm Zone

Constructing the Harnesses

Now that we have the test switches installed on the panel, it is time to start constructing the wiring harnesses which will connect the various input devices to the panel. The harnesses are made using a bundle of 22AWG solid wire and Molex connectors. I used four conductor telephone wire for the harness. The Molex connectors are designed to be used with an expensive crimp tool. I purchased a few extra pins for pennies and used needle nose pliers. One end of the LCD harness is shown below.

Figure 6-5. Molex Connector Pins

We do the same thing at the other end of the harness and then insert them into the shell. Slide the wire into the shell with the locking pin facing the locking slot until you hear a small 'click'.

Figure 6-6. LCD Display Harness

The figure above shows the assembled harness. Note that the connector shells are both keyed. The connector on the printed circuit board is keyed, but the one on the LCD display is not. Be sure to make both shells the same.

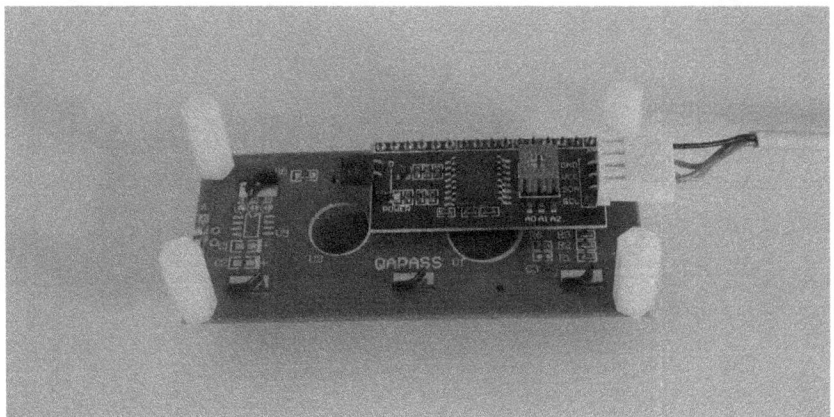

Figure 6-7. LCD Connector

The next device we will be wiring is the key switch. The key switch has three positions; left, right and center. The key is only removable in the center position. When turned left or right, the common terminal is connected to one of the other terminals. In my case, I used a black wire to indicate the common terminal. I try to use a common wiring color scheme. This will make trouble shooting a lot easier. I use red and black for power and ground and yellow and green for signals.

Additional keys can be ordered by quoting the serial number of the key, which is stamped on the front of the switch and on the key itself.

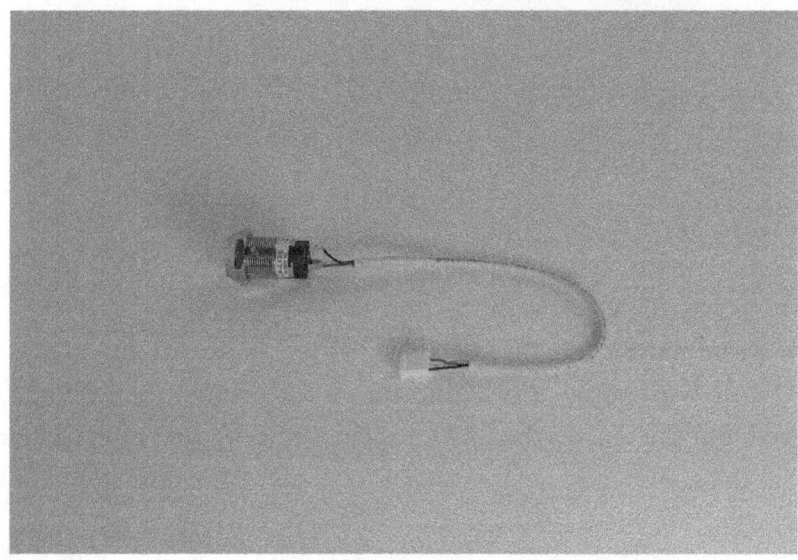

Figure 6-8. Key Switch Wiring

The pin numbers and functions are marked next to the pins on the rear of the switch. You will need a magnifying glass and a good light to read them.

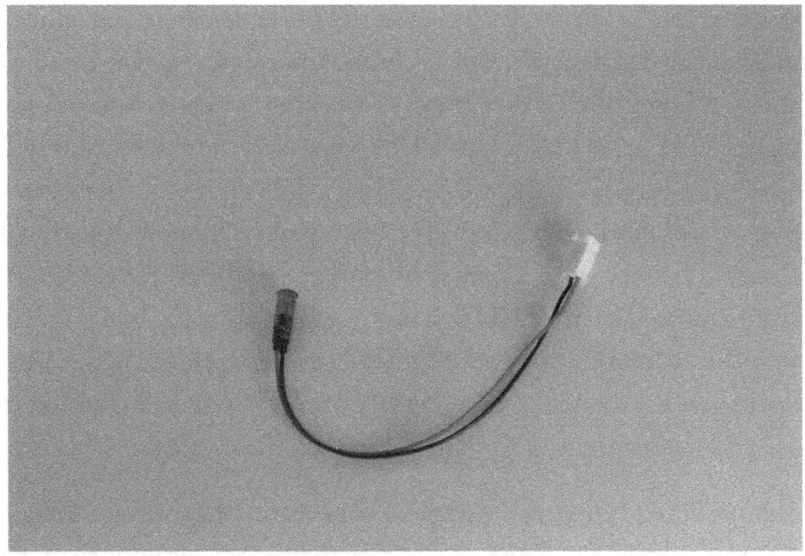

Figure 6-9. Alarm Indicator

A 12-volt panel lamp is used to simulate a siren. Note that the indicator I used is a polarized LED device. This is because the connector on the alarm panel is polarized to provide +12 volts to the siren. Most sirens require about 1 amp for power. The power cube, that I am using for testing, can only provide about 1 amp. For that reason, I am using a panel lamp. Also, panel lamps don't make any noise. This will become a factor during testing.

Testing the Harnesses

Now that we have the harnesses assembled, it is time to connect them to the alarm panel PCB. The key switch is connected to J1 (Key). There is a wire jumper connected to J2 (Tamper). This is where the tamper switch would be connected if one were installed on the panel case. This connector is shorted in order to prevent false alarms. The LCD display is connected to J3 (LCD) and the alarm indicator LED is connected to J4. Unsolder the battery pack from the PCB and remove the 12-volt power from the panel. Make sure that the Raspberry Pi is powered down. The alarm panel is connected to the Raspberry Pi via a 40-pin ribbon cable connected to J15. Once you have the ribbon cable connected, you can power up the Raspberry Pi and apply the 12-volt power. The LCD will display "System Idle" if everything is working properly. You can now plug the RFID card reader into a USB connector on the Raspberry Pi. You will also need speakers connected to the audio out jack. These can be USB powered speakers or externally power speakers for now. You can arm the system by swiping a 'Supervisor' card across the RFID reader. You will see a message on the LCD that says, "System armed in two minutes". This means that you have two minutes to exit the protected area. After two minutes has elapsed, the system will display "System Armed" on the LCD and will announce "System Armed" on the speakers.

Assuming you are running version 5.3 or greater of the software. Pressing the button attached to Zone 1 will cause the panel to announce that an "Entry detected in Zone 1" and the LCD will display a message that says "Entry Detected. You have 2 minutes". This means that you have two minutes to enter the building and disarm the alarm system. The system can be disarmed either by swiping either a normal RFID card or a supervisor card. You can also disarm the system by inserting the alarm key into the key switch and rotating it to the disarm position.

Pressing a button connected to any other zone will cause the panel to announce, "Alarm detected in Zone …" and the LCD will display a message "ALARM" "Zone …". The LED indicator connected to the Siren output will light, to simulate a siren activated.

Fire detection operates differently than the alarm portion of the panel. A fire alarm can be triggered whether the panel is armed or not. With the panel in the disarmed state, press one of the fire zone test buttons. The system will announce "Fire Detected" over the speakers and LCD will display "FIRE !" "Zone …" There are two fire zones, so the location of the fire is indicated. The fire alarm can be disabled using either the RFID card reader or the key switch.

When the panel is first powered up, it is in an 'Idle' state. This is indicated by the message on the LCD display. The panel must be initialized (Armed) using a supervisor RFID card. Once the system has been initialized, it can be armed or disarmed using the key switch. The system can also be disarmed using either a supervisor card or a normal user card. A supervisor card is the only RFID card that can arm the system.

You may recall that these cards were created in chapter 3, using the text editor that comes with the GUI to create two text files called "cards.txt" and "super.txt". The serial numbers of the 'supervisor' cards are stored in the super.txt file. These are the cards issued to management. The cards issued to normal workers have their serial numbers stored in the cards.txt file.

Wiring the Sensors

The first sensor we will be connecting is the door / window contactor. This is the simplest sensor we will be connecting to the panel. We start by trimming the leads of the end of line resistor as short as possible and then connecting one end of the resistor to one of the contacts. These devices are not powered, so we can remove the red and black wires. Trim the yellow and green wires as shown below. The device is not polarized, so you can solder either wire to the resistor. I use green to indicate the common terminal.

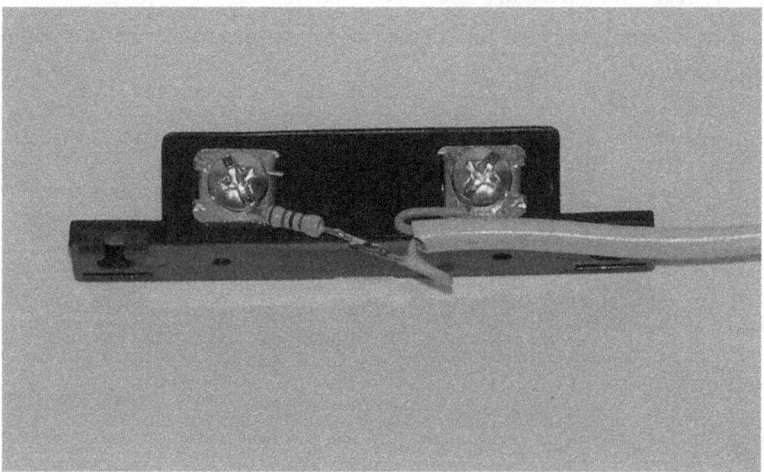

Figure 6-10. Step 1

Next, carefully fold the wires so that the alarm wire fits in the exit hole in the case.

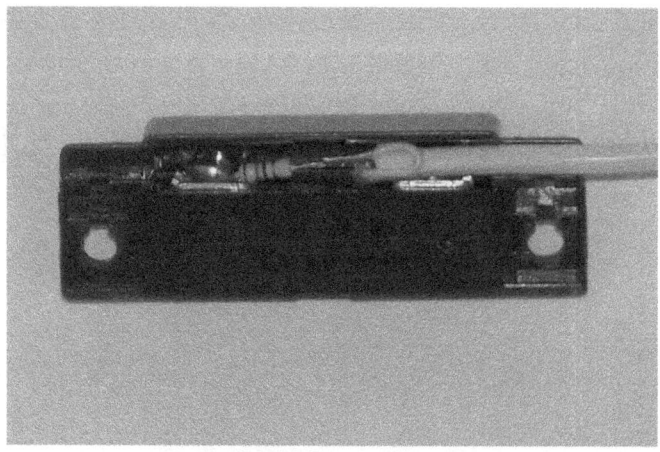

Figure 6-11. Step 2

Finally, we install the cover and make sure that the switch portion mates snugly with the magnet. If the sensor is mounted properly, you should read about 5.6K between the yellow and green wires with your ohmmeter. A broken wire will appear as an infinite resistance and a short circuit will appear to be zero ohms. A cold solder joint can appear as a much higher resistance. A careful inspection after step two will help eliminate these faults.

Figure 6-12. Step 3

If you read an infinite resistance instead of 5.6K, make sure that the contact portion of the device is properly seated against the magnet. It is the magnet that holds the contacts closed. If not properly installed the alarm system will think the window or door is open. The next device we will be connecting is a Passive Infrared Motion sensor. This device is a 4-wire device and is powered by 12 volts supplied by the panel.

Figure 6-13. PIR Motion Sensor

The red and black wires are connected to +12 volts and Ground on the zone terminal strip. The yellow and green wires are connected to the IN and COM terminals. To be consistent, I use the green wire as the common connection

Figure 6-14. Sensor Wiring

The figure above shows how the sensor is wired. Note the end of line resistor installed inside the case. The wiring has been pulled out of the case for clarity in the photo. In practice, the individual wires would be inside the case.

Figure 6-15. Tamper Switch Wiring

This figure shows the tamper switch wiring. Most alarm sensors have a built-in tamper switch which should be connected in series with the alarm contacts. The tamper switch is the small black switch to the right of the LED. With the cover snuggly in place, you should read about 5.6K ohms between the green and yellow wires.

The next device we will be discussing is the 4-wire smoke detector. This device is also powered by 12 volts from the panel.

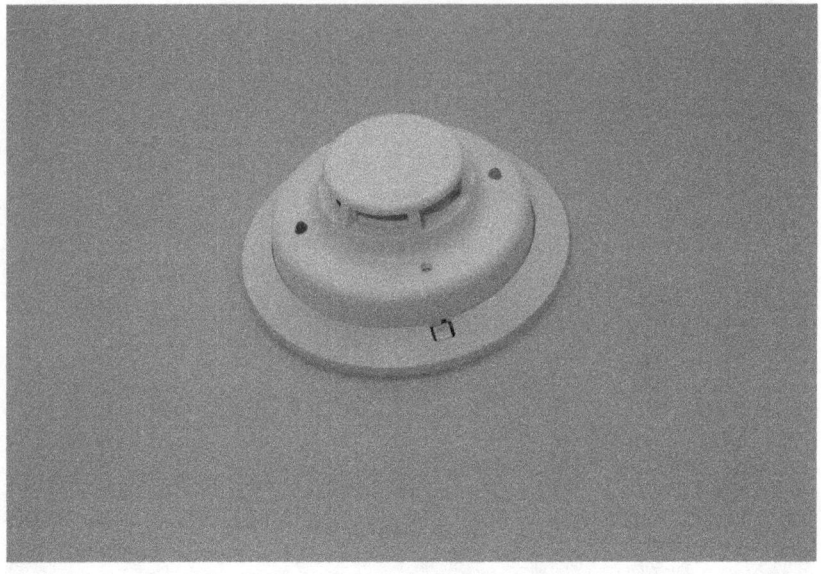

Figure 6-16. 4 Wire Smoke Detector

The device shown is an optical type of detector. The contacts on an ionization type detector will be similar to this one. There are two types of smoke detectors.

Optical detectors use infra-red LEDs to detect smoke inside a chamber which is closed to normal ambient light but allows smoke to enter. Smoke eventually obscures the light coming from the LED and detected by the infra-red detector. This triggers an alarm condition.

Ionization detectors use a small radioactive source to ionize the air inside a chamber, causing it to be conductive to electricity. Smoke causes the conductivity to rise until a certain point is reached. At this point the air is conductive enough to cause an alarm condition to be detected.

In both cases a relay is activated to signal an alarm. This relay is called a 'Form-C' because it has both normally open and normally closed contacts. We will be using the normally open contact.

This model of smoke detector has several contacts on the under side. The red and black wires are the 12 volt supply. The yellow and green wires are connected to the common and normally open contacts of the relay. This is to make the device compatible with the normally open fire zones of our alarm panel. Rate of Rise or ROR detectors may be connected in parallel with the smoke detector. They are a device with normally open contacts that close to indicate an alarm. That is why we are using the normally open contacts on the smoke detector.

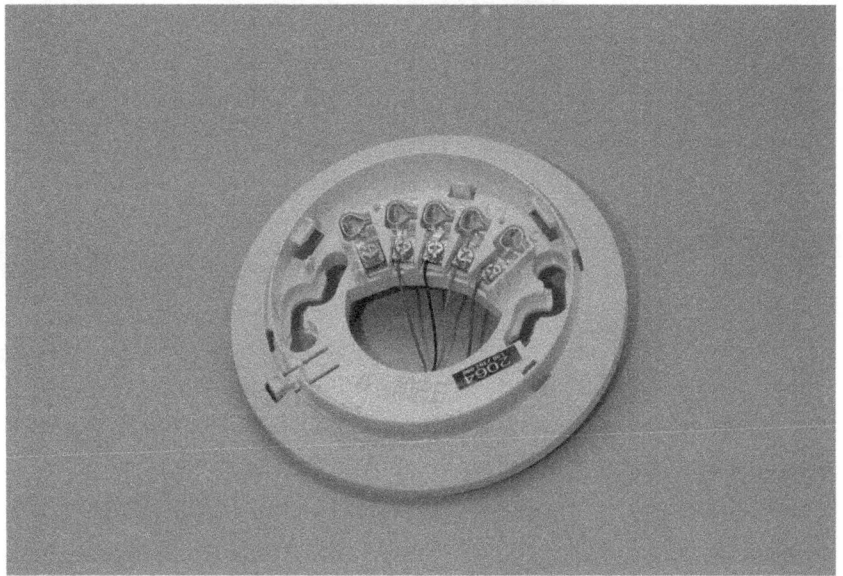

Figure 6-17. Smoke Detector Wiring

The device shown below is a combination rate of rise (ROR) and fixed temperature type fire detector.

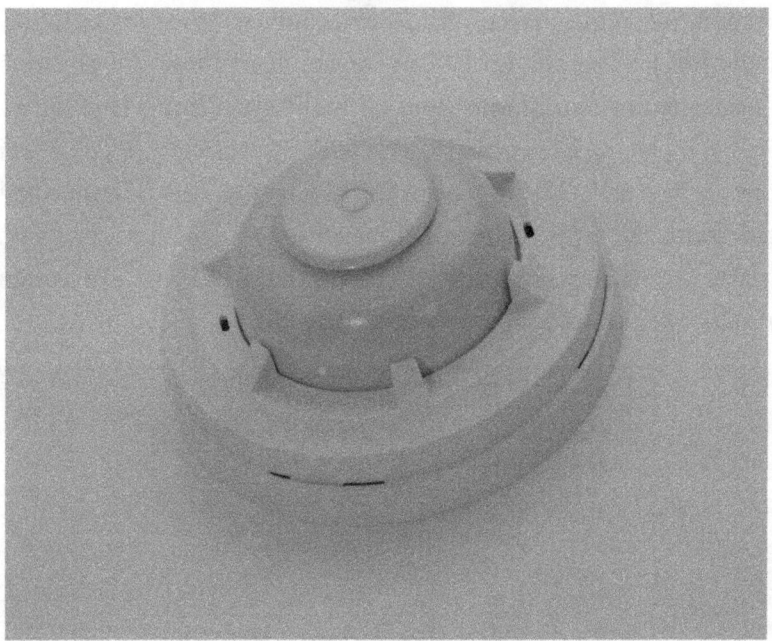

Figure 6-18. ROR Heat Detector

This type of detector is a 2-wire device because it has no electronics inside. It relies on a heat sensitive metallic switch which is normally open and closes when a fire is detected. In addition to the ROR element, there is also a small disk mounted on the device. This disk is mechanically connected to a set of contacts which also close if a preset temperature is reached. This temperature is normally marked on the case.

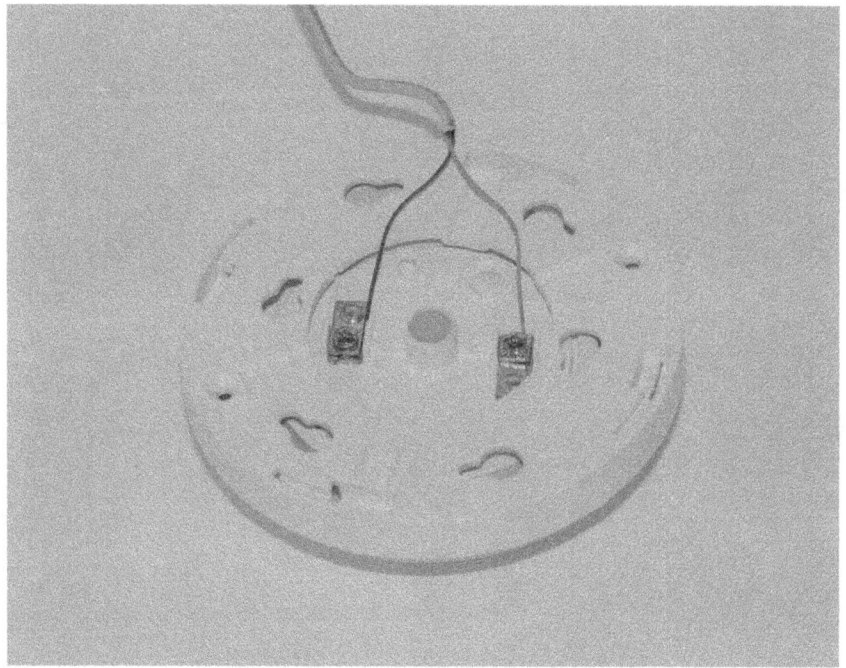

Figure 6-19. ROR Heat Sensor Wiring

The figure above shows the sensor wiring. As you can see there are only two normally open contacts. There are no end of line resistors installed on these fire detectors. This is because an antitamper feature is not used on our panel.

Smoke detectors and ROR detectors can be connected in parallel. This is because they are normally open contact devices. To do this you can either carefully insert two green and two yellow wires under the Common and IN terminals of the panel, or under the normally open terminals of either device.

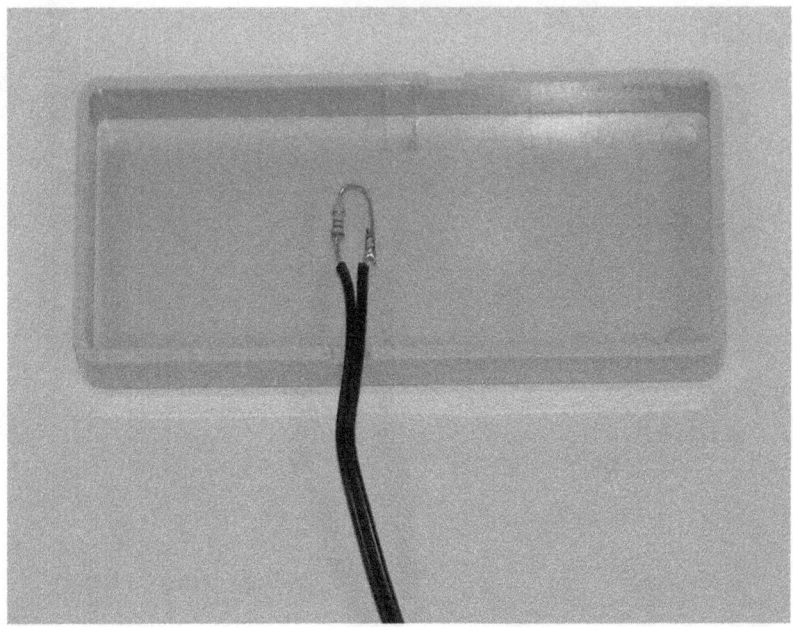

Figure 6-20. Simple Anti-theft device

The device shown above is not actually a sensor in the normal sense. It is a trick employed by many big box stores and other retailers. It is used to protect products with a handle or wheels like bicycles, lawn mowers etc. An end of line resistor is mounted inside a sealed plastic or metal box. The other end of the wire is threaded thru the handle of the lawn mower or the wheel of the bicycles. The wire is then connected to the alarm system with an inline plug and jack. An audio plug and jack will do fine. Two wire lamp cord makes an excellent wire for this application.

If a would-be thief disconnects the wire in order to steal the product, the connection to the end of line resistor is opened and an alarm sounds. If the same thief attempts to short out the cable, using a pin or some other object, the alarm panel will also detect this condition and sound the alarm. This simple device is basically a normally closed contact sensor with no moving parts.

Chapter 7:

Planning your Alarm System

In this chapter we will discuss how to plan an alarm system for your home and business. This chapter is provided for educational purposes only and should not be considered installation instructions for the alarm system described in this book. This is because no two houses or businesses are the same. What I am describing here is a typical installation.

There are two basic types of alarm systems: wired and wireless.

The wireless system is quick and easy to install. It is however easy for a professional burglar to jam the wireless signal that controls the alarm. There are numerous YouTube videos on the subject and the equipment required to carry out the attack costs less than $50USD on Amazon. For about twice that amount you can purchase a device capable of also jamming the cellular signal, so that the panel cannot contact the alarm company to tell them that it has been jammed.

The cost of a wired alarm system is considerably more up front. This is due to the labor involved in fishing wires thru walls and drilling holes to mount the sensors. The upside to a wired system is that there is no wireless alarm signal to jam. It is of course still possible to jam the cellular system or cut the phone line. That is why loud sirens were invented.

Our system is not monitored by an alarm system company so there is no one for the panel to call. Also, SMS messaging system protocols and frequencies vary from country to country. For that reason, an SMS option was not included in the design. I should point out that while this is a fun and educational project, there is no substitute for a live monitored professional alarm system. If you are protecting elderly or infirmed persons or a store that sells high end merchandise, then I would suggest that a live monitored wired system is the way to go. It is true that phone lines can be cut, and cellular links can be jammed, but a 110dBm siren will at the very least alert the neighbors and hopefully scare off a would-be thief. It might be advisable to tell your neighbors about your new alarm system and apologize in advance for any false alarms.

One other, somewhat controversial topic is stickers on your window the say "Protected by Acme Alarm Company". In my opinion, the sticker will deter the casual or amateur burglar and tell the pro what frequency to jam.

Planning is by far the most important part of the installation. Poor planning has caused more headaches than any other part of an alarm system installation project.

Step 1 – The Walk-about

- As the name suggests, take a walk around the property you will be protecting, both inside and out. Try to think like a bad guy.

- Ask yourself "If I wanted to break into this house / office / warehouse, how would I do it?"

- Take pictures if you think it will help. Make sketches of the inside and outside of the building so that you can plan where you are going to put your sensors. There are several free CAD programs that will help you with this. One popular program is called VISIO 2000. This software is over 20 years old and can be downloaded for free from various sites such as: "https://winworldpc.com/product/visio/2000" I used this software to draw the floor plans in this chapter.

- Here is a simple check list for you:

- How many doors are there?

- What type of doors are they? Garage doors? Human entry doors? Pet doors? (Seriously burglars have used children)

- How many windows are there? How many open?

- How many windows on the ground floor and how many on the second floor? (Burglars use portable ladders)

- Is there a hedge or a privacy fence. (Burglars love cedar hedges and privacy fences.)

- Once the burglar is inside, where can they go from there? (Main hallway, kitchen door, patio door)

- How many rooms are there that you will have to protect?

- Is everything on the same floor or is there more than one floor.

- What about special alarms? Panic alarm in the bedroom? Smoke alarm in the kitchen? Flood alarm in the basement? Heat detector in the garage?

Typical Four Bedroom House

The first system plan we will be discussing is a four-bedroom two story house with an attached garage.

Figure 7-1. Ground Floor

There are two entrances to the main floor. One via the mud room at the front of the house and one into the kitchen, from the garage. The main floor consists of a kitchen, a living room, and a dining room. There is a bay window in the living room and a picture window in the dining room.

The garage has two front and rear doors as well as the large garage door. The garage also has several small windows along one side.

The upstairs has four regular sized bedrooms and a study. There is also a master bedroom with an ensuite bathroom. Access to the bathroom is either from the master bedroom or the hallway.

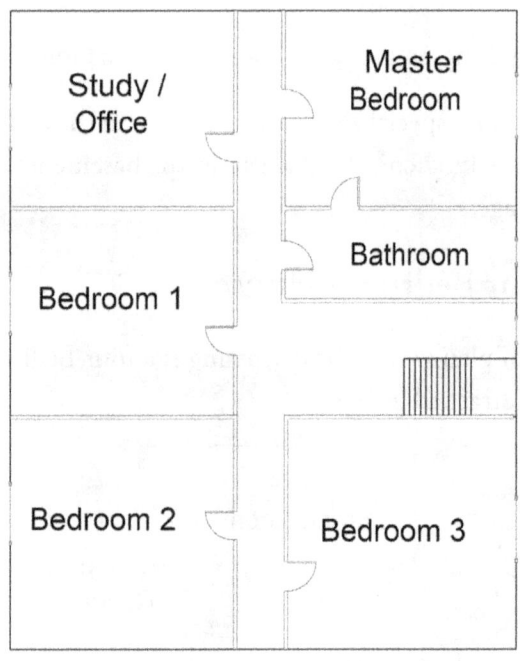

Figure 7-2. Top Floor

Now that we have our imaginary house designed, it is time to begin laying out our zones. We are protecting a two-story house which has bedroom windows on the top floor. I have decided not to alarm the windows. A good quality locking sash should do the job. There will be a smoke detector and a motion sensor covering the top floor landing. Most of the alarm zones are on the ground floor.

Zone 1 is a set of door contacts located on the interior mud room door. We will be using version 5.3 or greater of the software. This version has a two minute enter and exit delay. It uses Zone 1 as a special zone to accomplish this. The wall next to the door would be a good place to install the Arm/Disarm switch and a speaker, so that the user can hear the voice responses. Zone 2 is also a set of door contacts installed on the front door of the garage, next to the garage door. Large doors like the garage door can be difficult to alarm and are prone to false alarms on windy days.

Zone 3 is the door from the garage into the kitchen. These contacts are there to separate garage area from the living area, in the unlikely event that someone gains access to the garage. Zone 4 is window contacts on one of the garage windows. The other windows and back door are protected by a PIR motion sensor which is connected to Zone 7.

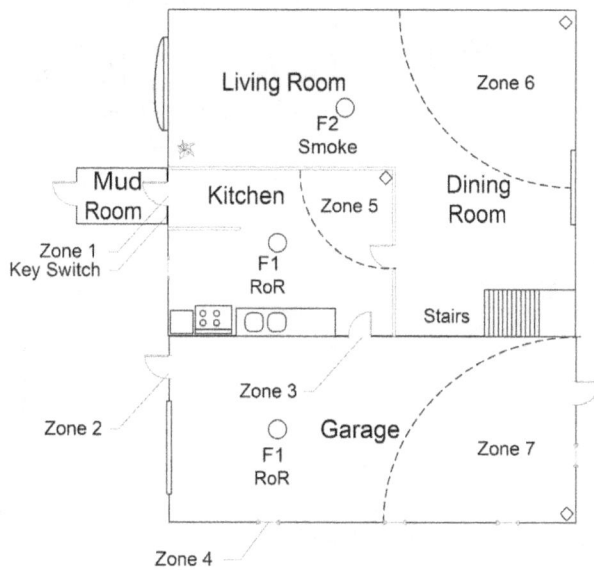

Figure 7-4. Ground Floor Sensors

Zone 5 is a PIR motion sensor which protects the kitchen area should anyone try to climb thru the kitchen window. It also covers the door to the dining room. The dining room and living room are covered by Zone 6. Zone 8 is a PIR sensor mounted at the top of the stairs to catch anyone trying to enter the upstairs area. There are two Fire Zones. Zone

F1 is comprised of two ROR (Rate of Rise) fire detectors connected in parallel. One is in the garage, the other in the kitchen area. The reason we use ROR detectors in these locations is so that burning toast in the kitchen and an idling vehicle in the garage does not set off an alarm. Zone F2 is a smoke detector zone with one detector located in the living room and one detector located in the upstairs hallway.

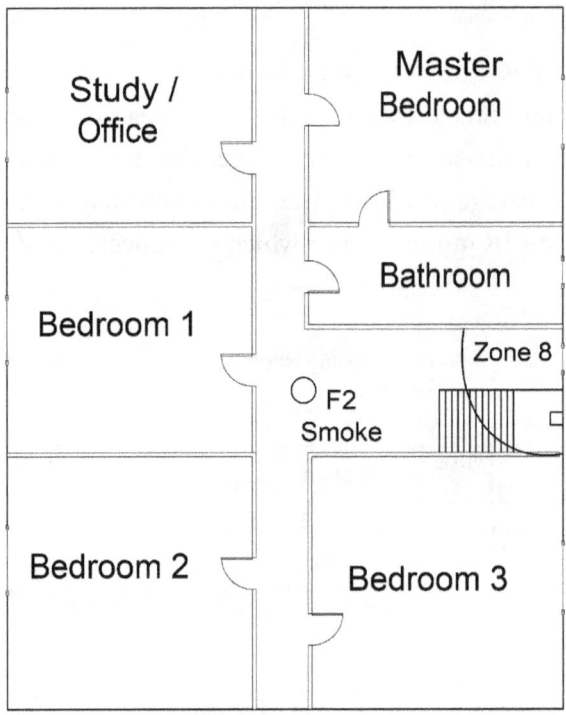

Figure 7-5. Top Floor Sensors

Smoke detector F2 is in the upstairs hallway at the top of the stairs. This detector is used to detect smoke coming from downstairs or one of the bedrooms. As shown above, the Zone 8 motion sensor is located at the top of the stairs. If the system is armed there is a chance that the motion sensor will detect someone walking down the hall. The study is perhaps a good place to have the alarm panel and the Raspberry Pi located. The two-minute exit delay should give the user plenty of time to reach either the front door or the kitchen door to the garage. If not, the delay can be easily changed in the software.

Typical Pub or Restaurant

Our next project is a commercial establishment like a small restaurant or your local pub. The main entrance is thru a set of double doors. There is also a staff entrance at the rear. There are several large windows in the main room and a bar and cash register near the door and front window. There is a door at the rear of the main room which provides access to the kitchen. A pantry is located behind the kitchen and is used to store food, alcohol, and perhaps other flammable goods. Across from the kitchen is the owner/manager's office. This is where we will install the alarm panel and the Raspberry Pi.

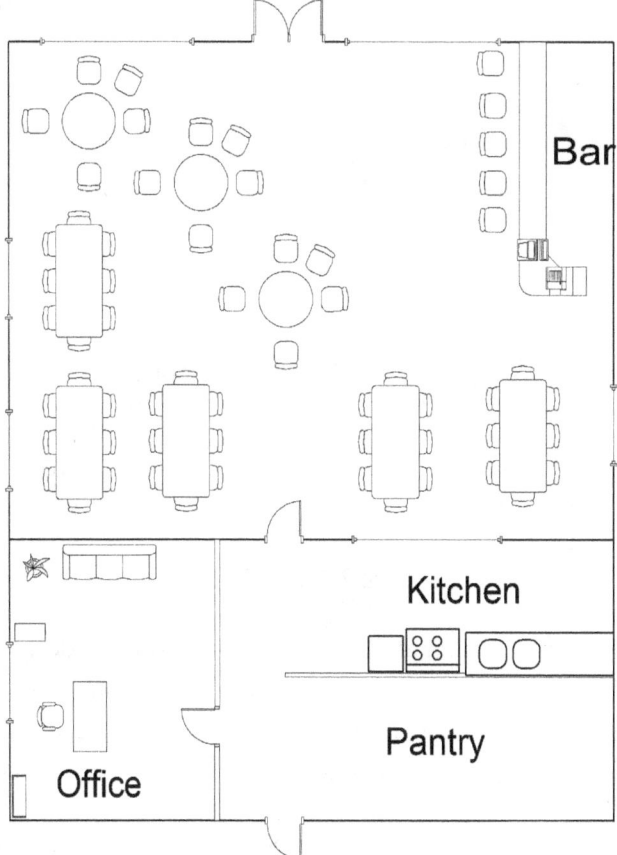

Figure 7-6. Restaurant / Pub

This layout will also use version 5.3 or greater of the software, because it has a two-minute entry and exit delay. The owner enters via Zone 1 and has two minutes to enter the office and disarm the alarm. An alternative to this is to mount the LCD display, the key switch, the RFID card reader, and a speaker in the hallway and run short cables thru the wall to the panel.

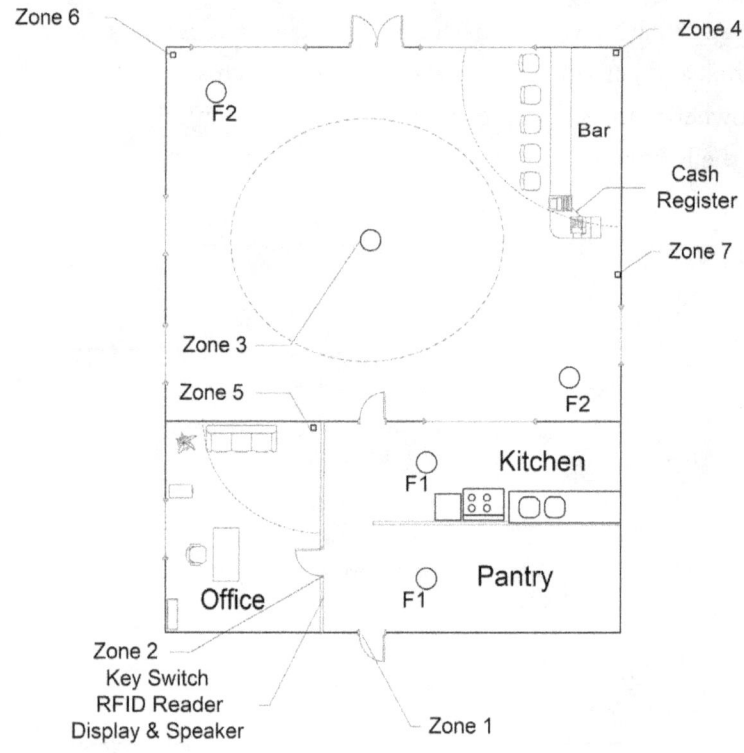

Figure 7-7. Pub Zones

Zone 1 is the rear entrance and has the two-minute entry delay associated with it. Zone 2 is the office door. This may seem redundant, given that we have a PIR motion sensor installed in the room as well. The Zone 5 motion sensor is there to catch persons entering thru the office window.

Zone 3 is a type of PIR detector we have not introduced until now. It is a 360-degree motion detector.

Figure 7-8. 360-degree Motion Detector

In the center is a Fresnel lens designed to give a full 360-degrees of coverage. If someone manages to slip by Zone 3; there is a standard motion sensor Zone 4 which protects the bar and cash register.

If this type of zone detector is not available in your area, then Zone 6 and Zone 7 can be installed on the wall as high up as possible. They are combination glass break and motion detectors and should have a clear line of site to the various windows in the dining area.

Figure 7-9. Typical Glass Break Sensor

The main front entrance is a set of double doors. These doors are not alarmed for the same reason that the garage door was not alarmed. People or wind rattling the front door will trigger a false alarm.

There are two fire zones. Zone F1 is two RoR fire detectors connected in parallel. This type of detector is used because smoke from the kitchen would falsely trigger a smoke alarm. The other detector is in the pantry and is used to detect smoldering fires. Zone F2 consists of two smoke detectors wired in parallel. These detectors are used to detect fires in the main dining area.

Commercial Office Space

Our next project is a typical office layout with suspended ceilings. This office is most likely located in an office building along with several other tenants. The layout consists of four open concept offices and three private offices, plus a board room. Each office has its own window but unless the office is located on the ground floor, glass break sensors are not required.

Figure 7-10. Office Layout

Entrance to the office is thru the main door, next to the reception area. This will be Zone 1 in the alarm system. The alarm panel controls can be located either in the hallway outside the office entrance, or inside near the reception desk. If you decide to mount the controls outside in the hallway, then there is no need for the entrance and exit delay. The raspberry Pi can be mounted under the counter in the reception area. Zone 1 is the main entrance. If you are using version 5.3 or greater of the software, the system will make an announcement when you enter or exit.

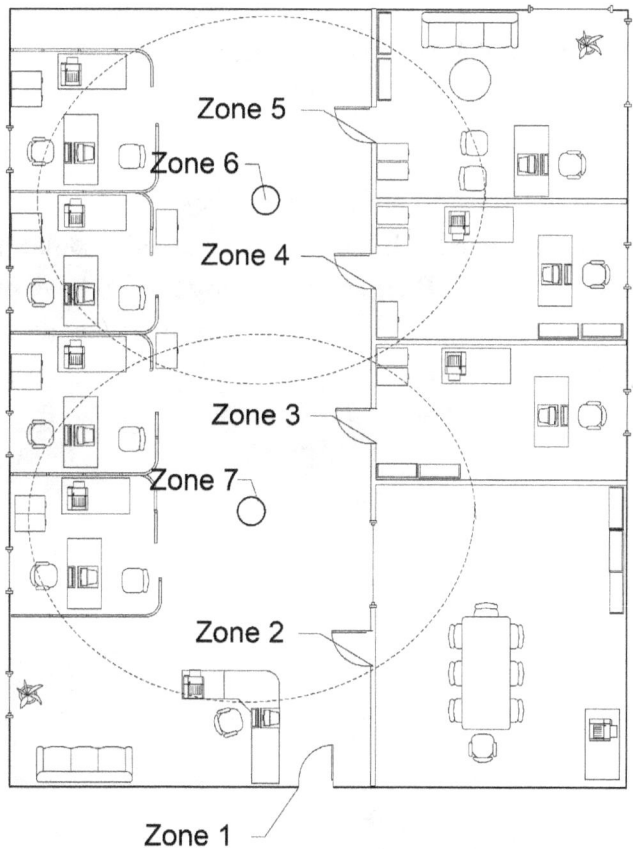

Figure 7-11. Office Zones

Zone 2 is a set of door contacts located inside the boardroom. Zone 3 and Zone 4 are also door contacts which are located inside the office. Zone 5 protects the office of the president or CEO. Zone 6 and Zone 7 are 360-degree PIR motion sensors. They can be 'standard' motion sensors. There is no need for the glass break option. These sensors are mounted to the ceiling in the hallway. Zone 7 protects the first two open concept offices and the reception area. Zone 6 protects the other two open concept offices. There is also some extra coverage of the private office doors. This may seem like it makes the door contacts on the private offices redundant. However, it is assumed that the occupant of the private offices is working on more sensitive work than the occupants of the open concept offices. There is also a smoke detector F1 located in the hallway. This can be a standard 4 wire smoke detector.

Laboratory

The project shown below is a research lab. For the sake of this design, we will assume that the lab is located inside an office building and is not on the ground floor. There are five alarm zones and two fire zones. This leaves three zones for specialized detectors such a CO detectors and various other fume detectors.

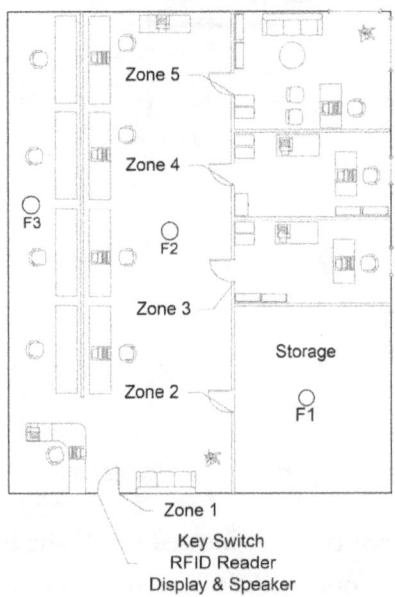

Figure 7-12. Laboratory Layout

Zone 1 is the main entrance to the lab. The RFID scanner, key switch, and LCD display can be mounted outside in the hallway. The Raspberry Pi and alarm panel can be mounted inside in the reception area. Zone 2 is the entrance to the storage room. This room contains a fire detector

F1. This can be a smoke detector or a ROR heat detector; depending on what is stored in the room. This could be flammable cleaners like acetone or alcohol. There is a door on the storage room, so installing a smoke detector to detect this sort of fire should not be a problem. A rate of rise detector would alert you to smoldering fires such as defective lithium batteries or soiled rags.

Zones 3, 4, and 5 protect the entrances to the private offices. There are two other fire zones which can be either smoke detectors or ROR type detectors. This will depend on what is being worked on in the lab and whether fume hoods are present. The following figure shows a combination CO (Carbon Monoxide) and smoke detector.

Figure 7-13. CO-Smoke Detector

This type of detector will be prone to false alarms if it is installed in an area where effective fume hoods are not installed. False alarms can be reduced using the type of sensor shown below. In both cases the detectors contain a chemical cartridge which must be inspected regularly and replaced if necessary, according to the manufacturer's specifications.

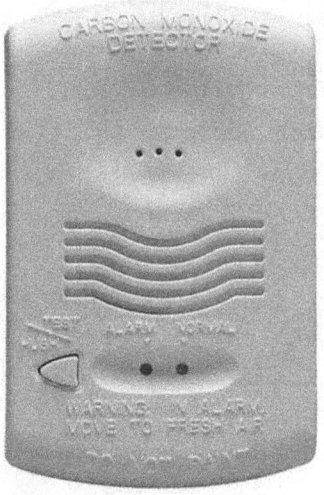

Figure 7-14. CO Detector

Chapter 8: Future Enhancements

The following chapter contains some ideas and suggestions for improving on your alarm system.

Running Alarm Program from Boot:

By now you should have an alarm panel with software that runs from the Genie IDE. This was Ok while we were debugging, but to make the system run 'headless' we have to make a few simple changes. First we have to edit our home .bashrc file. To do this open a terminal window and type: *sudo nano .bashrc*

This will display the text shown below. Scroll to the end of the file and enter the following lines. In this case I have panel52.py in my home directory, along with the cards.txt and super.txt files. Substitute the name of your home directory for "bill".

```
echo Running Alarm Panel 52
sudo python3 /home/bill/panel52
```

Press [Ctrl] - o to save the file, and then [Ctrl] – x to exit.

```
# enable programmable completion features (you don't need to enable
# this, if it's already enabled in /etc/bash.bashrc and /etc/profile
# sources /etc/bash.bashrc).
if ! shopt -oq posix; then
        if [ -f /usr/share/bash-completion/bash_completion ]; then
         . /usr/share/bash-completion/bash_completion
        elif [ -f /etc/bash_completion ]; then
         . /etc/bash_completion
        fi
fi
echo Running Alarm Panel 52
sudo python3 /home/bill/panel52.py
```

These changes will cause panel53.py or greater to be executed every time a terminal window is opened in the GUI. It will also run the program when we boot to the command line interface (CLI). This is how we run the system headless (without a keyboard, monitor, mouse).

The next thing we must do is to tell the Raspberry Pi to boot to the command line interface and automatically log in. We do this by navigating to the 'Preferences' tab and selecting 'Raspberry Pi Configuration'. Select **Boot: To CLI** and turn the **Auto login:** switch on.

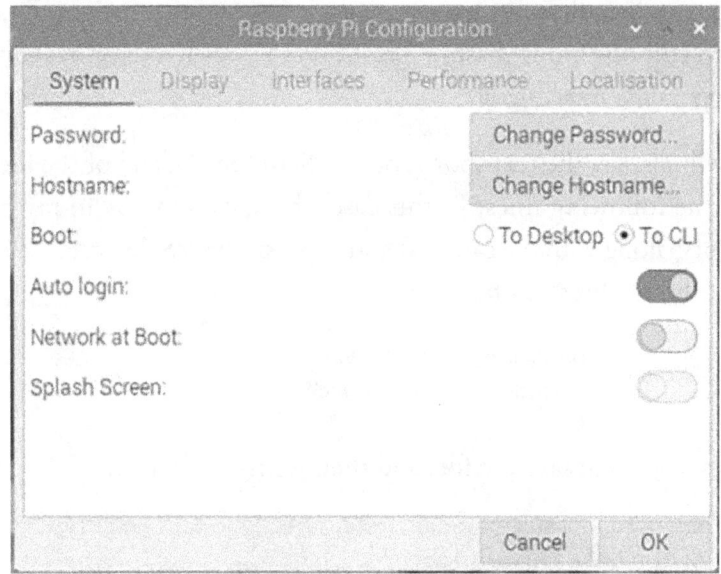

Figure 8-1. Raspberry Pi Configuration

Now if you reboot the system, you will have a headless system running. You may want to check that everything is running the way you like before disconnecting the monitor and other devices.

At this point the only devices that need to be connected to the Raspberry Pi are speakers and the USB card reader.

To get back to the GUI do the following:

The program will ask you if you want to reboot. Select <Yes> and the system will reboot back to your GUI.

>Enter: sudo raspi-config
>Select: System Options
>Select: Boot / Auto Login
>Select: Desktop Autologin Desktop GUI
>Select <Ok>

Water / Moisture Detector

The circuit shown below is for a single transistor water detector. None of the components are critical. Transistor Q1 can be almost any NPN transistor. Such as a 2N3904 or 2N2222.

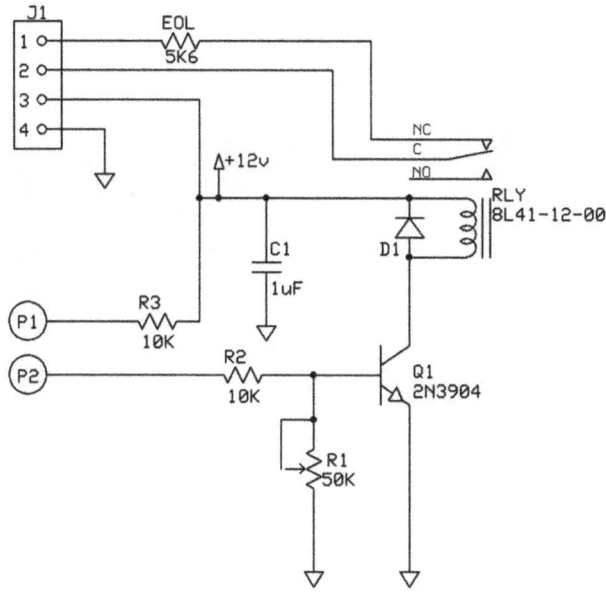

Figure 8-2. Water Detection Circuit

The relay can be a small 12 volt DIP relay. This sensor is a four-wire device designed to be powered by 12 volts from the panel. The sensor is calibrated by placing the probes in water and adjusting the 50kΩ potentiometer until you hear the relay 'click'. You may find that a multi turn potentiometer makes calibration easier. Connector J1 connects to the alarm system. Pin 1 of J1 is connected to the normally closed contact of the relay via a 5K6 ohm resistor. This resistor is used by the antitamper system. Pin 2 of J1 is connected to the common terminal of the relay. Pins 3 and 4 supply 12 volts to the board.

The probes are connected to the control PCB using thin stranded wire such as speaker wire, soldered to pads P1 and P2. When the circuit detects water the normally closed contact opens to signal an alarm, just like any other alarm system sensor. Probes can be constructed from scrap PCB material. You need two pieces of PCB material approximately one inch by two inches. Simply solder the stranded wire to the copper clad PCB material.

Darkness Detector

The circuit shown below is for a single transistor darkness detector. The detector is adjusted by covering R6 with black tape of some sort and adjusting R5 until the relay turns on. If you remove the tape the relay should turn off. Connector J2 is connected to an alarm system input. Pins 1 and 2 of the connector are connected to the common and normally closed (NC) contacts of the relay. A 5K6 ohm resistor is connected to pin 1. This is the End Of Line resistor and is used by the antitamper system. When the circuit senses darkness, the contacts open to signal an event just like any other alarm system sensor.

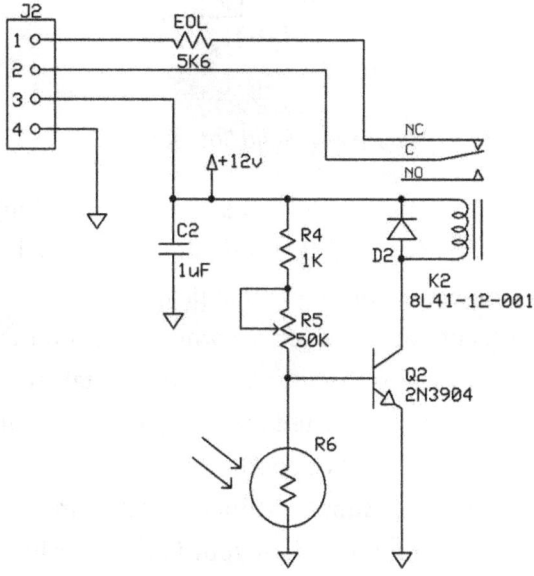

Figure 8-3. Darkness Detector

Adding a Door Solenoid:

There are two GPIO pins left on the Raspberry Pi. Future revisions of the PCB may include 'auxiliary' solid state relays which can be used to activate the solenoid of your choice. They will use the same IC as the siren relay. They will be 12 volt DC, 2 amp relays. If an expansion board is attached to the system, then eight solid state relay outputs will be available.

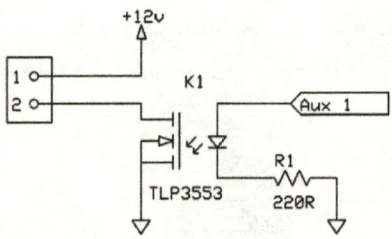

Figure 8-4. Solid Sate Relay

Solenoids come in many shapes, sizes, and applications. There are two basic types of devices, fail safe and fail secure. A fail safe device is unlocked when power is removed so that people can 'safely' enter or exit the area. A fail secure device is locked when the power is removed so that the area is protected or 'secure'. Care must be taken using this sort of device in an installation because the occupants of the area would be locked in should a fire or blackout occur. I would suggest that fail secure solenoids only be used for supply cabinets or other areas not occupied. In the event of a power failure, all of your fail safe devices will unlock, leaving the area unprotected. This is easily remedied using a standard dead bolt as a backup.

A typical fail secure door solenoid is shown below.

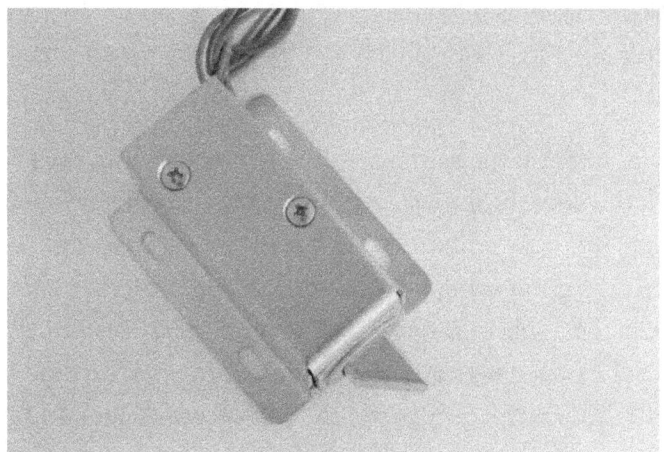

Figure 8-5. Fail Secure Door Solenoid

The photograph below shows the device with the cover removed. When power is applied to the coil it becomes an electromagnet, which pulls the latch mechanism into the coil. When power is removed, the spring returns the bolt to the locked state. The device I purchased has a spring which is extremely weak and might be easy to tamper with. For this reason and because it is a fail secure device, I would not recommend this device for any entrance or exit door applications.

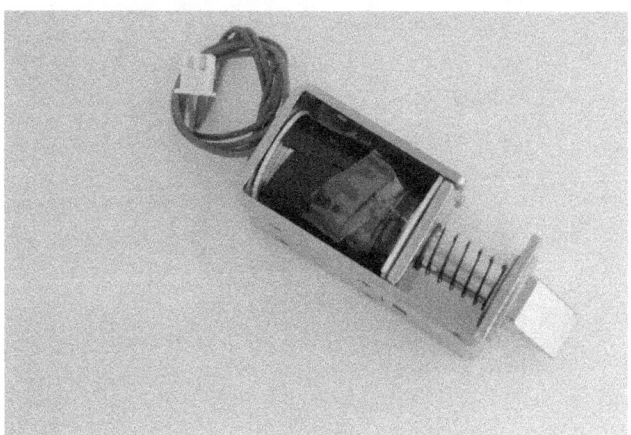

Figure 8-6. Cover Removed

The following information is provided for educational purposes only. I recommend that you contact a qualified lock smith for advice and installation. They will help you choose the lock set which best suits your situation.

In a typical scenario the user swipes the key card or uses the key switch to unlock the door. In the event of a power failure, the lock will default to 'fail safe' mode or unlocked. This will leave the protected area wide open. This is a good thing because the fire department or other EMT personnel will be able to access the area. If the fail safe strike plate shown in figure 8-12 is used with a simple door knob, a standard dead bolt lock can be installed on the door above the access control lock and used as a backup to secure the area once the area is safe.

The device shown below can be wired as either fail safe or fail secure. It could be used with a standard lock set. It is intended to replace the strike plate of the standard lock. A strike plate is installed in the frame of the door. The bolt extends into a hole behind the plate when the lock is in the 'locked' position.

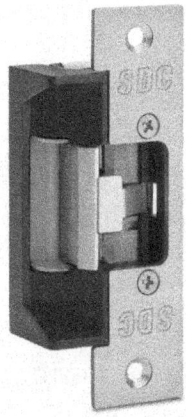

Figure 8-7. Fail Safe Strike Plate © Security Door Controls

There is a third type of solenoid lock which includes a key as a mechanical override. The lock shown below can be opened from the inside when the power fails. It also has a built-in clutch that helps prevent vandalism to the outside door handle. The door can also be opened from the outside using a key.

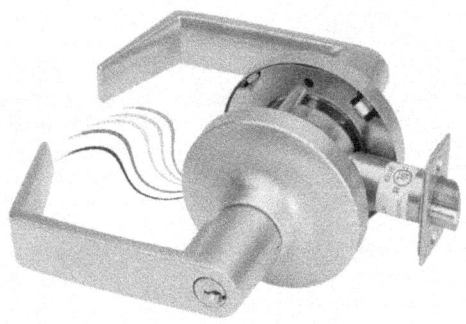

Figure 8-8. Solenoid Lock with Key © Security Door Controls

When the solenoid is activated will depend on your application. For example, the office scenario described in chapter 7. If the access controls are located outside of the protected area in the hallway, then disarming the system with a key card or a key can also unlock the door to zone 1.

It will also be necessary to modify the software so that the door solenoid is activated for about 1 minute when the system is disarmed. A sample of how to do this is shown below.

```
#Check Disarm Switch
        if GPIO.input(disarm):
                time.sleep(.1)
        else:

                lcd.clear()
                lcd.text(«System Disarmed»,1)
                arm_flg=0
                GPIO.output(alarm,GPIO.LOW) #Turn off siren


                #Activate door solenoid Code
                GPIO.output(aux1,GPIO.HIGH) #Activate door solenoid
                time.sleep(60) #Leave the door unlocked for 1 minute
                GPIO.output(aux1,GPIO.LOW) #Deactivate door solenoid
                ###############################################################

                os.popen('espeak -ven+f5 '"+ disarm_txt +
                '"--stdout | aplay 2> /dev/null').read()
```

Links:

CAD Software:

To the best of my knowledge, the following software is both legal and free. The Express PCB software allows you to price and order the PCB directly from the software.

https://winworldpc.com/product/visio/2000

https://www.expresspcb.com/pcb-cad-software/

Smoke Alarms:

https://www.safelincs.co.uk/smoke-alarm-types-ionisation-alarms-overview/

https://www.electronicsforu.com/technology-trends/smoke-detectors-fire-alarms-guide

https://medium.com/@chuanjerlim/confession-of-a-photoelectric-smoke-alarm-3be8bbd65af9

Solenoid Locks:

https://www.sciencedirect.com/topics/computer-science/electrified-mortise-lock

Suppliers:

Digikey is my main supplier of electronic components. They usually ship within 24 hours. The Bill of Materials contains both the manufacture's part number and the Digikey catalog number. Even if you don't purchase from them, Digikey has an extensive library of data sheets associated with each part.

https://www.digikey.de/en/products

https://www.digikey.co.uk/

https://www.digikey.com/

Security Door Controls is a worldwide supplier of Access Control hardware with sales offices in North America, England, and elsewhere.

https://www.sdcsecurity.com/Sales-Offices.htm

https://www.sdcsecurity.com/z7200-Series-Electric-Cylindrical-Lock.htm

https://www.sdcsecurity.com/electric-strikes.htm

Chapter 9: Adding more Inputs and Outputs

The following chapter contains some ideas and suggestions for improving your alarm system.

The MCP23017 I/O Expander:

The expander board is based on the MCP23017 port expander integrated circuit. The device comes with sixteen ports which can be configured as either inputs or outputs.

The I/O Expander Board:

The proposed expander board will have eight inputs and eight outputs. The inputs are identical to the ones on the main board and come with the same antitamper features. The eight outputs of the IC are connected to solid state, optically isolated relays. Each relay is capable of sinking twenty four volts at two amps, or forty-eight watts. This should be more than enough to drive mechanical relays, solenoids or relays.

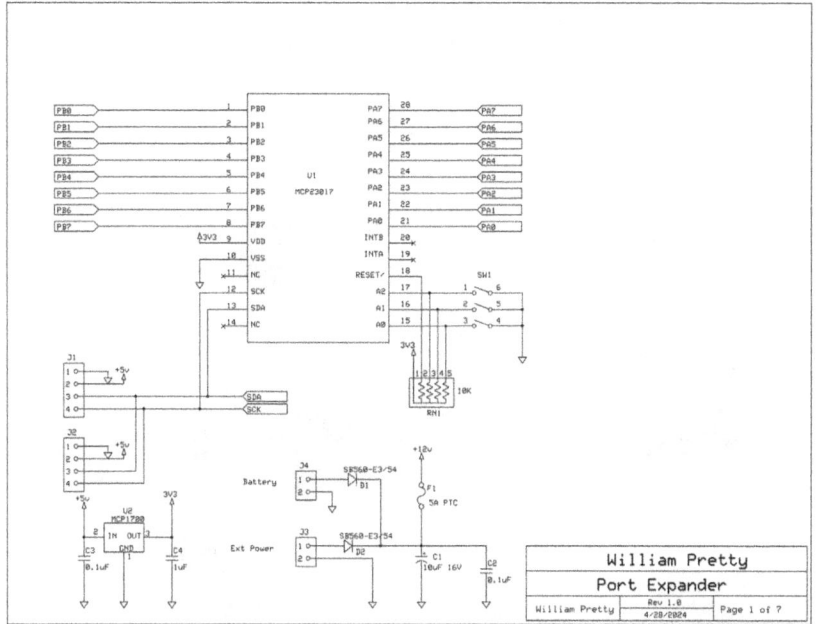

Figure 9-1. Port Expander IC

Figure 9-1 shows the connections to the MCP23017 port expander IC. Connectors J1 and J2 connect to the I2C bus which provides power and data to the expansion board. Connector J2 is provided as a loop through connector so that other expander boards can be connected to the system. Voltage regulator U2 is a low drop out 3.3 volt regulator which supplies power to U1 and the rest of the system. Connectors J3 and J4 provide 12 volt power to the board. The board is normally connected to a mains power supply which provides power for the board. The user can connect a 12 volt battery to J4 to supply power to the board in the event of a power failure. Three position dip switch SW1 and resistor network RN1 are used to set the address of the expander board. Closing a switch results in a logic 0 on the address line. In this system hex address x27 is used by the LCD display. For that reason, we have addresses hex x20 to hex x26 left over for our expander boards.

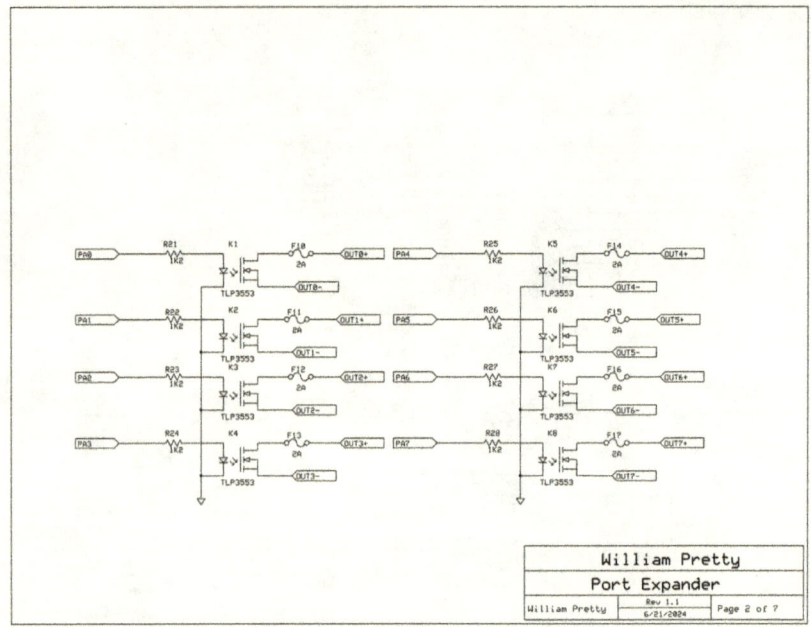

Figure 9-2. Solid State Relays

Figure 9-2 shows the eight solid state relays which are connected to Port A of the MCP23017. These are solid state relays so the correct polarity must be observed when connecting a DC load to these devices. Each output has its own ground connection (OUT0-) so that the ribbon cable ground connection does not have to carry several amps.

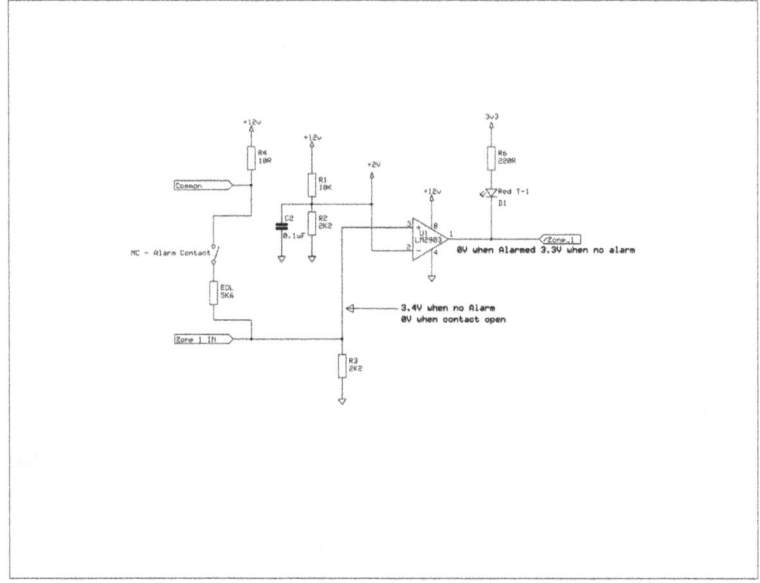

Figure 9-3. Alarm Inputs

Figure 9-3 shows a typical alarm sensor connection to the expander board. The alarm inputs on the expander board use exactly the same circuit as on the main board. The end of line resistor and the relay are located on the sensor device. Resistors R1 and R2 form a voltage divider which supplies approximately 2 volts to the negative input of the comparator U1. Resistors R3, R4 and the end of line resistor form a second voltage divider. In the normal condition the relay contacts are closed, and the voltage divider supplies 3.4 volts to the positive input of the comparator. During an alarm condition, the relay contact opens and resistor R3 pulls positive input to zero volts. When the positive input is higher than the reference to output is high. When the positive input is lower that the reference to output is low.

Trouble Shooting the System:

The alarm system software does not have to be running in order to perform these tests. Each zone input on both the main board and the expander boards has a red LED attached to its output. When an alarm condition occurs the output of the comparator goes low, and the LED turns on. A 5K6 resistor should be connected between the COM and the IN terminals of any zone inputs that are not in use. This simulates the end of line resistor and the LED associated with the zone should be off. If a normally closed push button switch is connected in series with the resistor, the switch can be used to simulate an alarm condition. A normally open push button switch can be connected across the same terminals of the fire zone inputs. Pressing the switch will make the comparator output go low and turn on the LED associated with the zone.

The address of the expander is set using the 3 position dip switch. Turning a switch on pulls the address bit low. With all of the switches closed the address of the first expander card will be Hex 20. This is because the most significant byte of the address is hard coded into the IC as a hex '2x' and cannot be changed by the user. Also, the address of the LCD is hard coded to Hex 27 and cannot be changed. These addresses can be checked using the *i2cdetect -y 1* Linux command. Some LCDs have a different hex address which may cause a conflict. Note that because of the default address of the LCD, only addresses Hex 20 through Hex 26 can be used by the expander cards.

Figure 9-4 i2cdetect Command

Hopefully this book and this chapter have inspired you to create your own high-tech alarm system.

I am sure that in the future others will take what I have started here and build on it. At least I hope so.

In the meantime, stay safe, have fun and happy building!

www.ingramcontent.com/pod-product-compliance
Lightning Source LLC
Chambersburg PA
CBHW020506030426
42337CB00011B/252